Annals of Mathematics Studies
Number 27

ANNALS OF MATHEMATICS STUDIES

Edited by Emil Artin and Marston Morse

1. Algebraic Theory of Numbers, *by* HERMANN WEYL
3. Consistency of the Continuum Hypothesis, *by* KURT GÖDEL
6. The Calculi of Lambda-Conversion, *by* ALONZO CHURCH
7. Finite Dimensional Vector Spaces, *by* PAUL R. HALMOS
10. Topics in Topology, *by* SOLOMON LEFSCHETZ
11. Introduction to Nonlinear Mechanics, *by* N. KRYLOFF *and* N. BOGOLIUBOFF
14. Lectures on Differential Equations, *by* SOLOMON LEFSCHETZ
15. Topological Methods in the Theory of Functions of a Complex Variable, *by* MARSTON MORSE
16. Transcendental Numbers, *by* CARL LUDWIG SIEGEL
17. Problème Général de la Stabilité du Mouvement, *by* M. A. LIAPOUNOFF
18. A Unified Theory of Special Functions, *by* C. A. TRUESDELL
19. Fourier Transforms, *by* S. BOCHNER *and* K. CHANDRASEKHARAN
20. Contributions to the Theory of Nonlinear Oscillations, *edited by* S. LEFSCHETZ
21. Functional Operators, Vol. I, *by* JOHN VON NEUMANN
22. Functional Operators, Vol. II, *by* JOHN VON NEUMANN
23. Existence Theorems in Partial Differential Equations, *by* DOROTHY L. BERNSTEIN
24. Contributions to the Theory of Games, *edited by* H. W. KUHN *and* A. W. TUCKER
25. Contributions to Fourier Analysis, *by* A. ZYGMUND, W. TRANSUE, M. MORSE, A. P. CALDERON, *and* S. BOCHNER
26. A Theory of Cross-Spaces, *by* ROBERT SCHATTEN
27. Isoperimetric Inequalities in Mathematical Physics, *by* G. PÓLYA *and* G. SZEGŐ

ISOPERIMETRIC INEQUALITIES IN MATHEMATICAL PHYSICS

By G. Pólya and G. Szegö

Princeton
Princeton University Press
1951

London: Geoffrey Cumberlege, Oxford University Press

Printed in the United States of America

PREFACE

The title of this book suggests its connection with a classical subject of mathematical research.

Two curves are called "isoperimetric" if their perimeters are equal. (We use here the term "curve" for a closed plane curve without double points.) The "isoperimetric problem" proposes to find, among all curves with a given perimeter, the curve with the largest area. The "isoperimetric theorem" states the solution of this problem: Of all curves with a given perimeter the circle has the maximum area. If we know the perimeter of a curve, but nothing else about the curve, we cannot find the exact value of its area. Yet the previous theorem yields some information, a definite bound, an inequality, the "isoperimetric inequality": the area is not larger than the area of the circle with the given perimeter. In reading this inequality in the opposite sense, we can restate the isoperimetric theorem in another form: Of all curves with a given area the circle has the minimum perimeter. Besides this great isoperimetric theorem, there are little isoperimetric theorems. For example, of all quadrilaterals with a given perimeter the square has the largest area. Moreover, of all quadrilaterals with a given area, the square has the shortest perimeter.

The Greeks knew the isoperimetric theorem, and had also some knowledge of its analogue in solid geometry. Pappus, in whose writings these results are preserved, attributes their discovery to Zenodorus. Two other names in the history of the isoperimetric theorem deserve our special attention.

The sphere is the most perfectly symmetrical solid. In changing a given solid into a sphere with the same volume, we diminish its surface area and give it an infinity of planes of symmetry. J. Steiner invented (1836) a geometric operation, called "symmetrization", which is, roughly speaking, a first step towards that radical change into a sphere: Steiner's symmetrization gives the solid at least one plane of symmetry, preserves its volume, and diminishes its surface area.

H. Minkowski added (1899) two new inequalities to the classical isoperimetric inequality between the volume and the surface area of a convex solid: one between the surface area and the average width of the solid, another between all three quantities: volume, surface area and average width. He showed also that the classical isoperimetric inequality is a consequence of these two new inequalities. (His investigation is restricted to convex solids.)

The history of our subject proper begins with B. de Saint-Venant. Investigating the torsion of elastic prisms, he observed (1856) that of all cross-sections with a given area, the circle has the maximum torsional rigidity. In fact, this observation appears in the work of Saint-Venant only as a conjecture, supported by convincing physical considerations and inductive (numerical) evidence, but without mathematical proof.

Lord Rayleigh observed (1877) that of all membranes with a given area, the circle has the minimum principal frequency. Again, this was only a conjecture; Rayleigh, however, supported it not only by inductive (numerical) evidence, but also by computing the principal frequency of almost circular membranes (the variation of the principal frequency) in a suggestive form. He stated, with less emphasis and less inductive or other evidence, two other similar conjectures in the same work:* Of all clamped plates with a given area, the circle has the minimum principal frequency; of all conducting plates with a given area, the circle has the minimum electrostatic capacity.

An important research of J. Hadamard (1908) on clamped plates had some analogies and connections with Rayleigh's second problem.

H. Poincaré stated (1903) with incomplete (variational) proof that of all solids with a given volume the sphere has the minimum electrostatic capacity.

T. Carleman (1918) proved the analogue of Poincaré's theorem in the plane, using conformal representation. G. Faber (1923) and E. Krahn (1924) found essentially the same proof for Rayleigh's conjecture concerning the membrane. Before them, R. Courant proved (1918) a weaker theorem; symmetrical membranes played an interesting rôle in his proof. G. Szegö gave the first complete proof for Poincaré's theorem on electrostatic capacity (1930) (Faber's paper contained an indication in this direction) and added a new inequality between electrostatic capacity, surface area and average width, of a convex solid (1931). G. Pólya and G. Szegö found (1945) that the electrostatic capacity of a solid is diminished by Steiner's symmetrization. This result led to several others; Saint-Venant's conjecture and the third of Rayleigh's three conjectures were completely proved, the proof of the remaining second essentially advanced, etc. All these results begin now to take their proper place within the framework of a general theory which the present book endeavors to outline.

---

* See No. 26 of the Bibliography, vol. 1, pp. 339-340 and p. 382, vol. 2, p. 179. This bibliography lists also the relevant works of Saint-Venant, Poincaré, and the others quoted in the sequel.

PREFACE

There are several interesting and important geometrical and physical quantities (set-functions, functionals) depending on the shape and size of a curve: the length of its perimeter; the area included; the moment of inertia, with respect to the centroid, of a homogeneous plate bounded by the curve; the torsional rigidity of an elastic beam the cross-section of which is bounded by the given curve; the principal frequency of a membrane of which the given curve is the rim; the principal frequency of a clamped plate of the same shape and size; the electrostatic capacity of this plate; and several other quantities. The isoperimetric inequality between the area and the perimeter was discovered first. Yet there are many similar inequalities connecting two or more of the quantities mentioned which have been discovered since or remain to be discovered. All these inequalities can be called, by extension, "isoperimetric inequalities". And there are analogous inequalities dealing with solids, or with couples of curves (condenser, hollow beam) or couples of surfaces, and so on. The present book is concerned with such inequalities.

These inequalities may be of some practical value. Of all triangles with a given area the equilateral triangle has the shortest perimeter. We may use this "little" isoperimetric theorem to find a lower estimate for the perimeter of a general triangle in terms of its area. As it is not difficult to compute the perimeter of a triangle, this estimate may be dismissed as a mere curiosity. Yet, of all triangular membranes with a given area the equilateral triangle has also the lowest principal frequency and we may use this theorem to find a lower estimate for the principal frequency of a general triangle in terms of its area. This estimate cannot be so lightly dismissed. First, it is simple and explicit; in fact, we know a simple explicit formula for the principal frequency of an equilateral triangle. Second, the estimate in question conveys information not easily obtainable otherwise; we do not know any explicit formula for the principal frequency of a general triangle. Third, the trend itself deserves attention: we estimate a not easily accessible physical quantity (the principal frequency) on the basis of easily accessible geometrical data (area, triangular shape). This is the trend of the present book: it aims at estimating physical quantities on the basis of geometrical data, less accessible quantities in terms of more accessible ones. The results hitherto obtained and discussed in this book allow already in some cases a fairly close estimate of physical quantities in which the engineer or the physicist may have a practical interest. And it seems possible to follow much further the road here opened.

The investigation here presented was undertaken in 1946 as a research project under the auspices of the Office of Naval Research. Several shorter reports and communications on the progress of our investigation have been submitted to this Office and to various meetings of the American Mathematical

Society and the Mathematical Association of America.* Under the title "Approximations and bounds for the electrostatic capacity and for similar physical quantities," a comprehensive paper was sent to the Transactions of the American Mathematical Society on September 20, 1948 and was accepted shortly afterwards. Difficulties of printing led to transferring the paper to the newly founded Memoirs of the Society, and eventually to suppressing the paper and publishing the present monograph instead, with the consent of all concerned.

The paper of September 1948 forms the bulk of the present volume, namely Chapters I-VIII. As this paper was widely circulated in the form of an O. N. R. report, repeatedly quoted by its original title, and several papers suggested by it have been printed already, we found it desirable to point out clearly its original contents. Therefore, all material additions and corrections made since September 1948 are enclosed in square brackets [ ]. Minor, merely formal alterations and additions which only expand references to publications prior to that date are not so marked, however. Moreover, we added new material in seven notes and Tables for some set-functions of plane domains. Note A was prepared by the first named author, the other notes, B to G, by the second named author. A first draft of the Tables, prepared by M. Aissen, H. Haegi, and J. L. Ullman under the supervision of the authors, has been submitted to the O. N. R. as a report on July 15, 1949. The notes survey the more important contributions to the subject obtained by the authors, sometimes in collaboration with others, up to the present date.

Here we observe that, as in the original paper of 1948, we leave certain not unimportant details to the physical intuition or the mathematical erudition of the reader. For instance, we do not insist on explicitly formulating and verifying the conditions under which the existence theorems for the boundary value problems considered have been proved. These conditions are important enough in themselves, but insistence on them would absorb too much time and effort and obscure the underlying facts and methods to which this investigation is devoted. We trust that in this form the whole book offers a fairly complete account of the present stage of an investigation that moves somewhat outside the usual channels, combines physical intuition and numerical evidence with analytical methods, and points to many still unsolved and interesting problems.

We wish to express our appreciation for the help and collaboration of several persons. H. Davenport, J. G. Herriot, Ch. Loewner, M. Schiffer, M. Shiffman and A. Weinstein participated in the research presented in the

---

* See, Bulletin American Math. Society, vol. 53 (1947), p. 65, p. 747, p. 1124, vol. 54 (1948) p. 667, vol. 55 (1949) p. 294, American Math. Monthly, vol. 54 (1947) pp. 201-206.

following. Also M. Aissen, H. Haegi, S. H. Lachenbruch, B. C. Meyer, W. J. Perry, Peter Szegö, B. M. Ullman, J. L. Ullman, and A. H. Van Tuyl contributed in various ways to this investigation.

The Office of Naval Research, and in particular Mina Rees and F. J. Weyl, took great interest in our work from the beginning. We acknowledge gratefully the friendly help of A. Erdélyi and A. W. Tucker in questions of publication. We wish to express our thanks to the Princeton University Press, in particular to Herbert S. Bailey, Jr., science editor, for the efficient handling of the printing. We are indebted to Peter Szegö for help in reading the proofs.

G. Pólya G. Szegö

Stanford University
January 1951

TABLE OF CONTENTS

CHAPTER I. DEFINITIONS, METHODS AND RESULTS

Dependence On Geometric Form

1.1. Space and plane. We shall examine geometrical and physical quantities (functionals) depending on the shape and size of a closed surface, or of a closed curve. In the order of treatment, figures in space will precede figures in a plane. We shall consider first a closed surface A surrounding a solid body B and various quantities connected with it, and pass afterwards to the case of a closed plane curve C without double points surrounding a plane domain D.

1.2. Space. The following quantities depend on the shape and size, but not on the location, of the solid B:

V the volume of B,

S the area of its surface,

C its electrostatic capacity (with respect to an infinitely large sphere).

If the solid is convex, we consider also

M the surface integral of mean curvature or (what Minkowski has shown to be the same) 2π times the average width of the solid.

For example, if our solid B is a sphere with radius r,

$$V = 4\pi r^3/3, \quad S = 4\pi r^2, \quad M = 4\pi r, \quad C = r.$$

It may be observed that the quantity C has some importance also outside the field of electrostatics. In fact, disregarding an obvious factor, we can interpret C also as a conductance, either thermal or electrical.

[We shall consider later still other physical quantities depending on the shape and size of a solid body B, especially its average virtual mass W_m and its average polarization P_m; see note G.]

1.3. Plane. The following quantities are connected with a plane domain D:

A the area of D,

L the length of its perimeter,

I_a its polar moment of inertia about the point a; the point a may lie anywhere in the plane, in or outside D. We conceive here D as covered with matter of uniform surface density 1. We consider an axis perpendicular to the plane of D and passing through the point a. The value of I_a, which is computed with respect to this axis, varies with the position of the point a.

I is the polar moment of inertia of D with respect to its center of gravity. Therefore, I is the minimum of I_a.

r_a is the inner radius of D with respect to the point a; the point a lies necessarily inside D. We map the interior of D conformally onto the interior of a circle so that the point a corresponds to the center of the circle and the linear magnification at the point a is equal to 1; the radius of the circle so obtained is r_a. The value of r_a varies with the point a.

$\dot{r}$ is the maximum inner radius. It is much less easy to characterize the points at which $\dot{r}$, the maximum of r_a, is attained than the point at which I, the minimum of I_a, can be found.

$\bar{r}$ is the outer radius of D. We map the complementary domain of D, exterior to C, conformally onto the exterior of a circle so that the points at infinity correspond and the linear magnification at infinity is equal to 1; the radius of the circle so obtained is $\bar{r}$.

C is the electrostatic capacity of D. We conceive here D as a cylindrical conductor of electricity with infinitesimal height, that is, as a metallic plate.

P is the torsional rigidity of D. We conceive here D as the cross-section of a uniform and isotropic elastic cylinder twisted around an axis perpendicular to D. The couple resisting such torsion is $\theta \mu P$. In this expression, θ is the twist or angle of rotation per unit length, μ is the modulus of shear, and P depends on the shape and size of the cross-section. The usual term for μP is torsional rigidity of the elastic cylinder. We introduce here for P the term "torsional rigidity of the cross-section." So defined, P is a purely geometrical quantity, independent of the units of mass and time; its dimension is that of the fourth power of a line. Its purely analytic definition will be given later, in 1.26.

Λ is the principal frequency of D. We conceive now D as the equilibrium position of a uniform and uniformly stretched elastic membrane fixed along the boundary C of D. The frequency of the gravest proper tone of this membrane is Λ. We make, however, Λ a purely geometric quantity by dropping a factor that depends on the physical nature of the membrane and has the dimension of a velocity. The analytical definition in 1.27 will leave no ambiguity. In fact, the case of the circle, mentioned in the next paragraph, might also suffice to remove the ambiguity.

If the domain D is the interior of a circle with radius r, all quantities mentioned can be computed:

$$A = \pi r^2 , \qquad L = 2\pi r , \qquad I = \pi r^4 / 2 ,$$

$$\dot{r} = r , \qquad \bar{r} = r ,$$

$$C = 2r/\pi , \qquad P = \pi r^4 / 2 , \qquad \Lambda = j/r .$$

We use here and in what follows the symbol j to denote the first positive root of the Bessel function $J_0(x)$,

(1) $$j = 2,4048\ldots \quad .$$

We may observe, that the quantities here listed are connected with further physical problems which we did not mention in outlining the above definitions. For example, we may conceive the closed curve C surrounding D as the cross-section of an infinite cylindrical conductor (of sheet metal). We may consider the electrostatic field in or outside this cylinder. The inner field is between the cylinder just described and a parallel infinite wire passing through the point a. The outer field is between the same cylinder and a parallel cylindrical conductor of which the cross-section is an infinitely large circle. The interior electric field is connected with the inner radius r_a, the exterior field with the outer radius $\bar{r}$ (details in 2.3). The outer radius $\bar{r}$ is also important in the study of the steady flow of an ideal fluid around the same infinite cylinder (wing of infinite span). If we are able to find the capacity C of the plate D, we can also solve an acoustical problem in which D is conceived as an aperture in a thin wall. Finally, the torsional rigidity P has several hydrodynamical interpretations of which we mention here just one: P is proportional to the discharge of a viscous fluid through a pipe of which the cross-section is D.

[We shall consider later still other physical quantities depending on the shape and size of a plane domain D, especially Λ_2, the principal frequency of a clamped plate, and Λ_3, a constant connected with the buckling of a plate. See sect. 6.6, 6.6 A and note F where the notation is different: we set $\Lambda = \Lambda_1$.]

1.4. Radii of a solid. Jointly with the solid B, we consider various spheres: a first sphere that has the same volume as B, another that has the same surface area, a third that has the same average width and a fourth that has the same capacity. The radii of these four spheres are

$$(3V/4\pi)^{1/3} = \bar{V} , \qquad (S/4\pi)^{1/2} = \bar{S} , \qquad M/4\pi = \bar{M} ,$$

and C, respectively. We shall call $\bar{V}$ the volume-radius of the solid, $\bar{S}$ its surface-radius and $\bar{M}$ its mean radius. (By placing a horizontal line over V, S and M, we derive from these quantities $\bar{V}$, $\bar{S}$ and $\bar{M}$, all of which have the dimension of a line.)

If the solid is not a sphere, we can hardly expect that any one of the three radii just defined will be exactly equal to the capacity C. Yet they may be approximately equal, and this will be certainly the case when the solid itself is almost spherical. Which one of the three radii, $\bar{V}$, $\bar{S}$ or $\bar{M}$, can be expected to yield the best approximation for the capacity? This question indicates the general trend of our investigation: We attempt to find relatively easily computable geometrical quantities suitable for approximating less easily computable physical quantities.

1.5. Radii of a plane domain. Jointly with a plane domain D, we consider various circles: a first circle having the same area, another having the same perimeter, and still others having the same polar moment of inertia, or capacity, or torsional rigidity, or principal frequency. The radii of these circles are

$$(A/\pi)^{1/2} = \bar{A}, \qquad L/2\pi = \bar{L}, \qquad (2I/\pi)^{1/4} = \bar{I},$$

$$\pi C/2 = \bar{C}, \qquad (2P/\pi)^{1/4} = \bar{P}, \qquad j/\Lambda = \bar{\Lambda},$$

respectively. Occasionally, we shall call $\bar{A}$ the "area-radius" and $\bar{L}$ the "perimeter-radius", and so on.

All these radii have the same dimension, that of a line. The dimensions of the quantities A, L, I, C, P and Λ, from which our radii have been derived, are various powers of the linear dimension.

1.6. Dependence on shape, size, and relative location. The quantities A, L, I, C, P, Λ, $\dot{r}$ and $\bar{r}$, introduced in 1.3, depend on shape and size of the domain D, or of its surrounding curve C, but not on its location. Yet, shape and size are not enough to specify I_a or r_a which depend also on the relative location of the domain D and the point a. On the other hand, products of powers such as L^2A^{-1} or PIA^{-4}, which contain only the 0th power of the linear dimension, depend on shape alone and are independent not only of location but also of size. Quantities of this nature, depending merely on the geometric form, deserve special attention.

In this connection, we mention also the so-called "virtual mass" of a solid moving through an ideal fluid; this quantity depends not only on shape and size of the solid, but also on the direction of its motion relative to a coordinate system attached to it. We quote this case here merely to illustrate another type of dependence. The virtual mass and the analogous concept of polarization will be discussed in note G.

Symmetrization

1.7. Space. Symmetrization is a geometric operation invented by Jacob Steiner.

Symmetrization with respect to the plane $\underline{P}$ changes the solid $\underline{B}$ into another solid $\underline{B}^*$ characterized as follows:

(I) $\underline{B}^*$ is symmetric with respect to $\underline{P}$.

(II) Any straight line perpendicular to $\underline{P}$ that intersects one of the solids $\underline{B}$ and $\underline{B}^*$ intersects also the other. Both intersections have the same length, and

(III) The intersection with $\underline{B}^*$ consists of just one line-segment (the intersection with $\underline{B}$ could consist of several segments). This segment is bisected by $\underline{P}$ and may degenerate into a point.

The plane $\underline{P}$, essential to the complete specification of the operation, is called the plane of symmetrization.

An example can be more conveniently given in the next section.

[The symmetrization here defined can be called, more specifically, symmetrization with respect to a plane, or Steiner symmetrization. Symmetrization with respect to a straight line, or Schwarz symmetrization, and symmetrization with respect to a point will be defined in sect. 7.1. Still other kinds of symmetrization (circular, spherical, continuous) will be introduced in notes A, B and C. Until the end of ch. 6, however, symmetrization means Steiner symmetrization, unless the contrary is explicitely stated.]

1.8. Plane. Symmetrization of a plane domain $\underline{D}$ with respect to a straight line l in its plane means symmetrization with respect to the plane $\underline{P}$ that passes through l and is perpendicular to $\underline{D}$. Therefore, $\underline{D}$ is changed into another plane domain $\underline{D}^*$. The domains $\underline{D}$ and $\underline{D}^*$ lie in the same plane, their intersections with any perpendicular to l are of the same length, and the intersection with $\underline{D}^*$ is a segment bisected by l. We call l the line of symmetrization.

For example, a semicircle of radius a symmetrized with respect to its bounding diameter is changed into an ellipse with semiaxes a and $a/2$.

By rotating the semicircle about the radius perpendicular to the bounding diameter, we obtain an example for symmetrization in space. A solid hemisphere of radius a symmetrized with respect to its bounding plane is changed into an oblate spheroid. This spheroid is described by an ellipse with semiaxes a and $a/2$ revolving about its minor axis.

1.9. Example. In order to observe the effects of symmetrization, we list some geometrical and physical quantities for the hemisphere and the oblate spheroid considered in the foregoing section. We list other quantities for the semicircle and the ellipse. The following tables have three columns each. The first column exhibits the symbol of a quantity, the second its value before symmetrization, the third thereafter. Where the exact value is not known, a blank is inserted.

	Hemisphere Radius a	Ellipsoid Semiaxes a, a, a/2
V	$2\pi a^3/3$	$2\pi a^3/3$
S	$3\pi a^2$	$2.760\pi a^2$
M	$3.57\pi a$	$3.42\pi a$
C	.845a	.827a

	Semicircle Radius a	Ellipse Semiaxes a, a/2
A	$\frac{1}{2}\pi a^2$	$\frac{1}{2}\pi a^2$
L	5.142a	4.844a
I	$.502a^4$	$.491a^4$
$\bar{r}$	.770a	.750a
$\dot{r}$	.6006a	.6060a
C	-	.464a
Λ	$3.832a^{-1}$	-
P	$.296a^4$	$.314a^4$

1.10. Theorem on symmetrization. The foregoing table illustrates the following general theorem:

> Symmetrization of a solid with respect to any plane leaves V unchanged and decreases S [and M] and C. Symmetrization of a plane domain with respect to any straight line in its plane leaves A unchanged, decreases L, I, $\bar{r}$, C and Λ, and increases P and $\dot{r}$.

What this theorem states about V and A is obvious. The results concerning S and L are due to Steiner; those about C, $\bar{r}$, $\dot{r}$, P and Λ have been proved by the authors (24, 21). See note A and ch. 7 where I and M are also discussed. We must add that we have used the words "decrease" and "increase" in the wide sense. Strictly speaking, we should have said "never increases" instead of "decreases" and "never decreases" instead of "increases."

The blank spaces in the table of 1.9 yield an opportunity to illustrate the practical use of our theorem. In fact, with its help, we can conclude from the table that the capacity of a semicircular plate of radius a is greater than $.464a$ and that the fundamental frequency of an elliptic membrane with semiaxes a and $a/2$ is less than $3.832a^{-1}$. These estimates are fairly sharp, and it may be quite troublesome to find better ones, especially for the capacity of the semicircle. Other cases in which the symmetrization yields useful numerical bounds will be discussed later, in 1.22.

Combining some elementary geometrical reasoning with our theorem on symmetrization, we can derive many curious results which would be very hard to obtain otherwise; see 7.4.

Inequalities

1.11. The role of inequalities. There are only a few cases (ellipsoid, lens, spindle, anchor-ring) in which a tolerably "explicit" formula has been found for the capacity, and the situation is similar with respect to principal frequency and torsional rigidity. There is no certain method for discovering "explicit" solutions even in simple cases. No such solution is known for the capacity of the cube or for the principal frequency of a rhombus.

When no explicit solution is known, we turn to approximations. An approximation involves a certain error. If we cannot estimate this error (which is often the case in practice) we can scarcely regard the approximation as satisfactory. A reliable approximation should yield definite bounds. Realizing this, we shall see more clearly the interest of the following sections where we intend to discuss general inequalities, capable of yielding numerical bounds in any proposed particular case, and fairly close bounds in some particular cases.

1.12. Inequalities derived from symmetrization. The sphere is symmetrical with respect to any plane passing through its center. Repeated symmetrization with respect to a suitable chosen infinite sequence of planes changes any solid, ultimately, into a sphere of equal volume. Similarly, any plane domain can be transformed into a circle of equal area by symmetrization with respect to an appropriate sequence of straight lines. Whereas V and A remain unchanged, the other quantites considered in our theorem of sect. 1.10 are changed continually in a definite direction by repeated symmetrization and go over ultimately into the corresponding quantities for the sphere (or circle); these latter, however, can be expressed in terms of the volume of the sphere (area of the circle). In this way, we are led to the following inequalities for solids:

$$(1) \qquad S^3 \geqq 36\,\pi\, V^2\,, \qquad\qquad C^3 \geqq 3V/\,4\,\pi\,,$$

and to the following ones for plane domains:

$$L^2 \geqq 4\,\pi\, A\,, \qquad\qquad I \geqq A^2/\,2\,\pi\,,$$

$$(2) \qquad \bar{r}^2 \geqq A/\pi \geqq \dot{r}^2\,,$$

$$C^2 \geqq 4A/\,\pi^3\,, \qquad \Lambda^2 \geqq j^2\,\pi/\,A\,, \qquad P \leqq A^2/\,2\,\pi\,,$$

In relations (1), equality is attained in the case of the sphere, in (2) in the case of the circle. The first inequality under (2) is the isoperimetric inequality properly so called. By extension, all the inequalities (1) and (2) may be called isoperimetric inequalities.

For the proof of the second inequality under (1) (Poincaré's theorem) see 3.2, for the second (obvious) inequality under (2) see 7.2, for remarks on the proof of the other inequalities see note A.

1.13. <u>Convex solids</u>. The inequalities 1.12 (1), to which we were led by symmetrization, deal with three quantities V, S and C, and apply to any kind of solids. Passing to the special class of convex solids, we have one more quantity to deal with, Minkowski's M mentioned in 1.2, and a few more inequalities.

We begin by comparing the classical inequalities

$$(1) \qquad (3V/\,4\,\pi)^{1/3} \leqq (S/\,4\,\pi)^{1/2} \leqq M/\,4\pi$$

with

$$(2) \qquad (3V/\,4\,\pi)^{1/3} \leqq C \leqq M/\,4\,\pi\,.$$

In each of these four relations equality is attained in the case of the sphere, and in this case only.

Using the notation introduced in 1.4, we can rewrite (1) and (2) in the form:

$$\bar{V} \leqq \bar{S} \leqq \bar{M}\,,$$

$$\bar{V} \leqq C \leqq \bar{M}\,,$$

or still shorter as follows:

$$(3) \qquad \bar{V} \leqq \begin{matrix} \bar{S} \\ \\ C \end{matrix} \leqq \bar{M}\,.$$

We mention one more fact. For certain solids $\bar{S}$ is greater than C, for others C is greater than $\bar{S}$; neither $\bar{S} \leq C$ nor $\bar{S} \geq C$ is generally true. We take (3) as expressing also the fact just mentioned.

The inequalities (1) and (2), or the more condensed (3), suggest a certain analogy between $\bar{S}$ and C which will be further disclosed by later discussions.

Those inequalities under (1) and (2) which do not involve M have been given already, in slightly different form, under 1.12 (1). The inequality between S and M is due to Minkowski who gave, in fact, a considerably sharper result. He proved the two parallel inequalities

$$(4) \qquad 4\pi S \leq M^2 , \qquad 3MV \leq S^2 .$$

Only the first was mentioned under (1). By eliminating M from (4), we can derive the classical isoperimetric theorem dealing with V and S (the first inequality under 1.12 (1)) which appears, therefore, as decomposed into two more fundamental inequalities.

The second inequality under (2) is due to Szegö who proved, in fact, the considerably sharper theorem (31):

$$(5) \qquad C \leq \frac{M}{4\pi} \frac{2\varepsilon}{\log \frac{1+\varepsilon}{1-\varepsilon}} \qquad \text{where} \qquad \varepsilon^2 = 1 - \frac{4\pi S}{M^2}$$

By using the terminology introduced in 1.4, we can express the contents of (5) in intuitive terms: <u>The capacity of a convex solid is never larger than the capacity of a prolate spheroid, the major semi-axis of which is the mean radius of the solid and the minor semi-axis the surface-radius.</u> We shall give a new derivation for (5) in 3.4.

An estimate of the capacity of an ellipsoid (see 24, (1.7)) leads to the following inequality for C which is less sharp than (5), but sharper than the second half of (2):

$$(6) \qquad C \leq (\bar{M} + 2\bar{S})/3 .$$

1.14. <u>Solids of revolution</u>. We consider here briefly another subclass of solids, the solids of revolution. A plane passing through the axis of such a solid intersects it in a plane domain which we call the <u>meridian section</u>. A meridian section has necessarily an axis of symmetry.

If C denotes the capacity of the solid of revolution and $\bar{r}$ the outer radius of its meridian section, we have

$$(1) \qquad C \leq \frac{4}{\pi} \bar{r} .$$

Equality in (1) is attained only in the limiting case when the solid of revolution becomes a circular disk and its meridian section a straight line-segment; see 3.7.

1.15. Plane domains. The inequalities 1.12 (2), obtained by symmetrization, compare the area A with L, I, $\bar{r}$, $\dot{r}$, C, Λ and P. Are there further inequalities between these latter? The best answer that we can give to this question for the time being is contained in the following system of inequalities:

$$\dot{r}^4 \leqq 2P/\pi \leqq (A/\pi)^2 \leqq 2I/\pi \leqq \bar{r}^4 \leqq (L/2\pi)^4 ,$$

(1)

$$\dot{r}^2 \leqq (j/\Lambda)^2 \leqq A/\pi \leqq (\pi C/2)^2 .$$

By using the notation introduced in 1.5, we rewrite (1) in the following equivalent, but more concise form:

$$\dot{r} \leqq \bar{P} \leqq \bar{A} \leqq \bar{I} \leqq \bar{r} \leqq \bar{L} ,$$

(2)

$$\dot{r} \leqq \bar{\Lambda} \leqq \bar{A} \leqq \bar{C} .$$

In all relations listed under (1) and (2), equality is attained in the case of the circle, and only in this case, with one exception. The inequality connecting I and $\bar{r}$ (the last but one in the first line, in (1) or (2)) goes over into an equality if, and only if, the curve is a "lemniscate of the 2nd order"; the meaning of this term and the proof will be explained in 5.19.

The inequalities between $\dot{r}$ and P, $\dot{r}$ and Λ, I and $\bar{r}$ are, to our knowledge, completely new.

The classical isoperimetric inequality between A and L results in our system (1) from three other inequalities (between A and I, I and $\bar{r}$, $\bar{r}$ and L) and appears, therefore, as decomposed into more fundamental inequalities.

There are reasons to suspect that a more comprehensive system of inequalities is valid, which we can express using the notation of 1.13 (3) as follows:

(3)
$$\dot{r} \leqq \bar{\Lambda} \leqq \bar{P} \leqq \bar{A} \leqq \frac{\bar{I}}{\bar{C}} \leqq \bar{r} \leqq \bar{L} .$$

Yet, what we definitely know to be true of this conjectured system is contained already in (1) or (2), except for one fact: For some plane domains $\bar{I}$ is greater than $\bar{C}$, for others, conversely, $\bar{C}$ greater than $\bar{I}$. The inequalities $\bar{\Lambda} \leqq \bar{P}$ and $\bar{C} \leqq \bar{r}$ are conjectural.

1.16. Further inequalities. Our knowledge about further inequalities between the quantities considered is rather incomplete. Still, we feel that we should mention certain partial results and even some conjectures. We shall give an exhaustive table later (1.21). Here we restrict ourselves to a few comments on selected examples which illustrate the kind of questions encountered in this sort of investigation. Discussing these examples, we shall also find opportunity to prepare for the conventional concise notation in which the table of sect. 1.21 will be presented.

1.17. An approximate formula due to de Saint-Venant. The approximate formula for torsional rigidity

$$P \sim (2\pi)^{-2} A^4 I^{-1} \tag{1}$$

is due to de Saint-Venant. (He preferred sometimes the numerical factor 1/40 to $(2\pi)^{-2}$, which makes a difference of less than 1.5%.) As it stands, formula (1) has a certain advantage: its two sides become exactly equal whenever the cross-section is an ellipse. On the basis of the theorem of sect. 1.10 we can say something more in favor of (1): both sides vary in the same direction when the cross-section is symmetrized. Yet, if we wish to judge clearly the merits of (1), we must investigate the bounds between which the quantity PIA^{-4} varies.

The first result of this investigation is disappointing. As suitable examples show (see p. 250) the lower bound of PIA^{-4} is 0 and its upper bound ∞: the ratio of P to the expression that allegedly approximates it is, in fact, completely unrestricted.

Further investigation, however, speaks more in favor of (1). We find that PIA^{-4} remains between finite positive bounds provided that the cross-section D is convex; see 5.11. That is to say, there are two positive constants, k and k', such that

$$k A^4 I^{-1} < P < k' A^4 I^{-1} \tag{2}$$

for any convex cross-section. We do not know the best possible values of k and k'. If it should turn out (which is likely) that k'/k is not much superior to 1, this would not only justify the approximate formula (1), but would also render it much more useful.

1.18. Capacity and surface area. An interesting approximate formula for the capacity C of a solid was proposed by A. Russell*:

$$C \sim (S/4\pi)^{1/2} . \tag{1}$$

* The eighth Kelvin lecture. Some aspects of Lord Kelvin's life and work, The Journal of the Institution of Electrical Engineers, vol. 55 (1917) pp. 1-17.

In words, <u>the capacity is approximately equal to the surface-radius</u>. We can say a few things in favor of (1). The two sides become exactly equal in the case of the sphere, both vary in the same direction when the solid is symmetrized, and both are contained between the same limits according to 1.13 (1) and 1.13 (2).

Yet, we find that the bounds for $CS^{-1/2}$ are 0 and ∞. That is, the right hand side of (1) turns out to be an arbitrarily large multiple of the intended left hand side for certain solids, and an arbitrarily small fraction for other solids. Even if we restrict the investigation to convex solids, the situation does not change completely. We obtain a one-sided result: For convex solids, the lower bound of $CS^{-1/2}$ is positive but the upper bound remains ∞. We have a double inequality of the form

(2) $$k < C\, S^{-1/2} < \infty$$

of which the right hand side is not too informative in concrete cases. The bound k in (2) has, of course, nothing to do with the k in 1.17 (2); in both formulas k stands for some fixed positive number.

In the case of formula (2) of the present section, however, we can at least guess with some likelihood the best value of k. We conjecture that the ratio $C/S^{1/2}$ becomes a minimum for the circular disk. If we had a proof for this conjecture, we could replace (2) by the remarkable inequality

$$C^2 \geqq 2\pi^{-3} S$$

which, of course, assumes that the solid is convex.

1.19. <u>Other examples</u>. Imitating de Saint-Venant's formula 1.17 (1), we propose the following approximate expression for the principal frequency:

(1) $$\Lambda^2 \sim 12\pi^2 I A^{-3} .$$

In discussing (1), we can repeat with little change whatever we said of 1.17 (1). The two sides of (1) become exactly equal when the membrane has the shape of a rectangle or (which is curious) of an equilateral triangle. Both sides of (1) are changed in the same direction by symmetrization. Yet, for membranes of suitable shape, the right hand side turns out to be an arbitrarily large or small multiple of the left hand side. This cannot happen, however, for convex membranes; for these, the ratio of the two sides of (1) remains between finite positive bounds.

In the conventional terse notation of the table in sect. 1.21, the facts about 1.17 (1) and formula (1) of this section are summarized as follows:

$$0 < k < P\,I\,A^{-4} < k' < \infty,$$

$$0 < k < \Lambda^{-2}\,I\,A^{-3} < k' < \infty;$$

see entries 19 and 20. The bounds closer to the variable quantity in question hold for convex cross-sections; they are (unspecified) finite positive numbers (k denotes a positive finite quantity, but not necessarily the same quantity in different formulas, and the same holds for k'). The bounds further away from the variable quantity hold for general unrestricted solids; the upper bound is ∞, the lower bound is 0.

In similar notation, the facts discussed in 1.18 are summarized as follows:

$$0 < k < C^2\,S^{-1} < \infty;$$

see entry 3 in 1.21. The upper bound is the same for convex and for general solids. We stated a conjecture about the value of k in 1.18; the formula here given expresses what we know for certain at the present moment.

A similar situation is summarized by:

$$? < k < P\,\Lambda^4 < \infty;$$

entry 14 of the table in 1.21 is different in form, but equivalent in fact. This expresses the following state of knowledge: We know that, both for convex and for non-convex, domains the upper bound is ∞. We know that the lower bound is positive for convex figures. We do not know anything non-trivial about the lower bound for unrestricted figures: it could be 0, or positive but less than the bound for convex figures, or equal to this latter. (We suspect that the last alternative is true and that the lower bound is attained in case of the circle.)

1.20. Other data. There is a difference of a practical sort between the various quantities considered. We can regard V, S, M, A, L, I_a and I, which depend on the evaluation of definite integrals, as relatively elementary. Yet r_a, $\dot{r}$, $\bar{r}$, P, Λ and C are definitely less elementary; they depend on the solution of boundary value problems related to partial differential equations in two or three dimensions. It is often appropriate to regard the more elementary quantities as known or given, the less elementary as unknown. Our inequalities, for example 1.13 (2), help to estimate the unknown (less elementary) quantities in terms of the data (of the more elementary quantities).

We could, of course, think of collecting more data, of introducing further geometrical characteristics of solids or curves. We naturally hope that with more data in hand we shall be able to estimate the unknowns more closely. The table in 1.21 gives a survey of our attempts in this direction (entries 22 - 37). Here we restrict ourselves to a few comments on selected examples.

(a) The density of electric charge on a conductor varies from point to point. This variation has been linked to the variation of the curvature. Therefore, it is natural to seek some estimate of the capacity C in terms of R_1 and R_2, the principal radii of curvature of the conducting surface at a variable point. In order to simplify our formulas, we introduce the principal curvatures

$$\kappa_1 = 1/R_1 , \qquad \kappa_2 = 1/R_2 .$$

With this notation, we have for any smooth convex solid

$$(1) \qquad \iint \frac{\kappa_1 - \kappa_2}{\log \kappa_1 - \log \kappa_2} \, dS \leqq 4\pi C \leqq \iint \tfrac{1}{2} (\kappa_1 + \kappa_2) \, dS ;$$

dS is the element of surface, and the integrals are extended over the whole surface of the solid. The second integral represents Minkowski's M; in fact, $(\kappa_1 + \kappa_2)/2$ is the mean curvature. The second inequality under (1) differs only in notation from the second inequality under 1.13 (2). Equality in (1) is attained only in the case of the sphere.

(b) The first integral in (1) is a geometric quantity attached to the convex solid; we do not know of any further question in which this quantity would play a role. Here is another geometric quantity, related to a convex curve C which arises in several problems:

$$(2) \qquad \int h^{-1} ds = B_a .$$

In order to define (2), we select a point a inside C. Let h denote the length of the perpendicular drawn from a to the tangent at a variable point of C where ds is the line-element, and extend the integral (2) over the whole curve C. The value of B_a varies with the point a. We let B denote the minimum of B_a; thus, B is a geometric quantity (functional) depending on the shape of a convex curve C: Being of dimension 0 (a mere number), B is independent of the size of C. [There is just one point a inside C at which B_a attains its minimum. This point lies, therefore, on any axis of symmetry that C may have, and coincides with the center of symmetry of C if C has one. See M. Aissen, 1, where it is also shown that B is diminished by symmetrization.]

We can estimate P and Λ, the torsional rigidity and the principal frequency of the domain surrounded by C, in terms of its area A and of the constant B as follows:

$$(3) \qquad P \geqq A^2 B^{-1} ,$$

$$(4) \qquad \Lambda^2 \leqq j^2 B(2A)^{-1} ;$$

we used here the notation 1.3 (1).

There is another case in which the integral (2) is useful. Consider a curve $\underline{C}'$, similar to $\underline{C}$ and similarly located so that the point a that we selected inside $\underline{C}$ in computing (2) is the center of similitude. Let $\underline{C}'$ surround $\underline{C}$, and let L' and L denote the lengths of their perimeters, respectively. We consider the condenser formed by two infinite conducting cylinders with cross-sections $\underline{C}$ and $\underline{C}'$, respectively. Let c denote the capacity of this condenser per unit height. (Distinguish carefully c from C as defined in 1.2.) Then we have the inequality

$$(5) \qquad c \leqq \frac{B_a}{4\pi \log (L'/L)} ;$$

L and L' denote here, as we said, the lengths of the perimeters of $\underline{C}$ and $\underline{C}'$, respectively. If $\underline{C}$ is a circle, equality is attained in (3), (4) and (5).

An integral analogous to (2) can be formed in space. See 3.5.

(c) Finally, we wish to emphasize entries 26 and 27 of the table. They give both an upper and a lower estimate for the capacity of a solid of revolution in terms of the mapping function of the exterior of the meridian section. The upper estimate is less simple but generally sharper than that given by 1.14 (1).

1.21. Table. In order to condense the information offered, we make use of various conventions and abbreviations; we begin by explaining them.

Entries 1 to 21 of the following table are concerned with the quantities V, S, M, C, A, L, I, $\dot{r}$, $\bar{r}$, P and Λ introduced in 1.2 and 1.3. Each entry estimates some combination of these quantities. These combinations have the dimension of a pure number; that is, they are not affected by the choice of the units. The quantities estimated are included between certain bounds. Entries 4 to 6 are concerned with convex solids only. The other entries give information both on convex and unrestricted figures. If, on a certain side of the estimated quantity, only one bound is given, it refers both to convex and to unrestricted figures. If there are two different bounds, the one closer to the quantity estimated is for convex figures, and the other (further) bound for unrestricted figures. The name printed next to the bound specifies the particular figure for which the bound is attained, or in the neighborhood of which the bound is approximated.

The last column refers to sections of the present paper where fuller information may be found, or occasionally to the Bibliography. In some cases it indicates also the authors to whom the result is due, explains the notations used, offers a conjecture, etc.

Symbols, Terms, Abbreviations

$\leq$ is used if the bound is known and is attained.

$<$ is used if the bound is not known, or if it is known but not attained.

! follows the name of a particular figure if the bound is attained only for it.

? expresses lack of information.

k and k' denote positive numbers, not necessarily the same in different formulas.

plate is a convex plane domain, considered as extreme case of a convex solid.

disk is a circular plate, limiting case of an oblate spheroid.

needle is a line-segment, considered as extreme case of a convex plane domain, or of a convex solid.

conj. : conjectured.

pr. : proved.

For terms not explained here consult the sections of this book quoted in the last column.

Entries 22 to 37 involve various quantities not introduced in 1.2 or 1.3, and follow the above conventions more loosely. In this part of the table, it may be more often necessary to refer to the sections quoted in the last column.

Example. Entry 13 summarizes the following facts:

$\dot{r}\Lambda$ has the precise upper bound j (value 2.4048..., see 1.3 (1)). This bound is valid for all plane domains, convex or not, and is attained for the circle alone.

$\dot{r}\Lambda$ has a positive lower bound for convex domains. The last column offers the conjecture that this lower bound is 2, the value attained in the extreme case of a strip, that is, the infinite convex domain between two parallels.

Nothing is asserted about the lower bound for unrestricted plane domains ; not even a conjecture is offered.

Other examples explaining the notation have been discussed in 1.19.

SPACE

Solids

1.	plate	$0 \leqq \frac{V^2}{S^3} \leqq \frac{1}{36\pi}$	sphere!	
2.	plate	$0 \leqq \frac{V}{C^3} \leqq \frac{4\pi}{3}$	sphere!	Poincaré, Faber; pr. Szegö (30)
3.		$0 < k < \frac{C^2}{S} < \infty$	needle	7.6; conj. $k=2\pi^{-3}$, disk
		Convex Solids		
4.	needle!	$0 \leqq \frac{S}{M^2} \leqq \frac{1}{4\pi}$	sphere!	Minkowski
5.	needle!	$0 \leqq \frac{C}{M} \leqq \frac{1}{4\pi}$	sphere!	3.4, Szegö (31)
6.	plate	$0 \leqq \frac{MV}{S^2} \leqq \frac{1}{3}$	Kappenkörper	Minkowski
		Solids of Revolution		
7.	needle!	$0 \leqq \frac{C}{\bar{r}} \leqq \frac{4}{\pi}$	disk!	3.7, $\bar{r}$ of meridian section
		PLANE		
8.	needle	$0 < \frac{\dot{r}^4}{P} \leqq \frac{2}{\pi}$	circle!	5.7
9.	needle	$0 < \frac{P}{A^2} \leqq \frac{1}{2\pi}$	circle!	5.17, conj. Saint-Venant (27)
10.	needle	$0 < \frac{A^2}{I} \leqq 2\pi$	circle!	7.2
11.	needle	$0 \leqq \frac{I}{\bar{r}^4} \leqq \frac{\pi}{2}$	lemniscate 2 ord.!	5.19
12.		$0 < k < \frac{\bar{r}}{L} \leqq \frac{1}{2\pi}$	circle!	5.11; conj. $k=1/8$, needle
13.		$? < k < \frac{r}{\Lambda^{-1}} \leqq j$	circle!	5.8, 5.11; conj. $k=2$, strip

PLANE (Continued)

14.	strip $0 \leqq \frac{\Lambda^{-4}}{P} < k < ?$		5.11; conj. $k=2\pi^{-1}j^{-4}$ (circle) also bound for non-convex
15.	strip $0 \leqq \frac{\Lambda^{-2}}{A} \leqq \frac{1}{\pi j^2}$	circle!	conj. Rayleigh, pr. Faber, Krahn (26, 5, 14)
16.	needle $0 < \frac{A}{C^2} \leqq \frac{\pi^3}{4}$	circle!	conj. Rayleigh (26), pr. (24)
17.	needle $0 \leqq \frac{C}{\bar{r}} < k$		conj. $k=2/\pi$, circle (24)
18.	needle $0 < \frac{I}{C^4} < k < ?$		5.11
19.	$0 < k < \frac{P}{A^4 I^{-1}} < k' < \infty$		1.17, 5.11
20.	$0 < k < \frac{\Lambda^{-2}}{A^3 I^{-1}} < k' < \infty$		1.19, 5.11
21.	$? < k < \frac{\Lambda^2 P}{A} < k'$		5.4, 5.11

SPACE

Solids

22.	$C \leqq D/2$	sphere!	D diameter, conj. Maxwell (15), pr. Szegö (31)
23.	$C \geqq \frac{2}{\pi}\left(\frac{P}{\pi}\right)^{1/2}$	disk	7.6; P maximum orthogonal projection

Convex Solids

24.	$C \leqq \frac{M}{4\pi} \frac{2\varepsilon}{\log \frac{1+\varepsilon}{1-\varepsilon}}$	sphere!	3.4, Szegö (31), $\varepsilon^2 = 1 - 4\pi SM^{-2}$
25.	$C \geqq \frac{1}{4\pi}\iint \frac{1/R_1 - 1/R_2}{\log(1/R_1) - \log(1/R_2)} dS$	sphere	3.4; R_1, R_2 principal radii of curvature

Solids of Revolution

26.	$C \leqq \left[2 \int_{\bar{r}}^{\infty} \left(\int_0^{\pi} r u d\theta \right)^{-1} dr \right]^{-1}$		3.6, u real part of outer mapping function
27.	$C \geqq \int_0^{\pi} \left[2 \int_{\bar{r}}^{\infty} (ru)^{-1} dr \right]^{-1} d\theta$		3.6
28.	$C + C' < k\bar{r}$		3.7, conj. k=2, sphere

Condensers

29.	$\frac{S}{4\pi d} < C < \frac{S}{4\pi d} + \frac{M}{4\pi}$		parallel coatings at distance d (24)
30.	$C < \left(\frac{S'}{S}\right)^{1/2} \iint \frac{dS}{4\pi d}$		3.5, coatings similar and similarly located, d distance between corresponding tangent planes

PLANE

31.	$\bar{r} \leqq D/2$	circle!	D diameter
32.	$\bar{r} \geqq p/4$	needle!	7.6; p max. orth. proj.
33.	$\Lambda^2 \leqq j^2 B (2A)^{-1}$	circle	1.20, 5.6
34.	$P \geqq A^2 B^{-1}$	circle	1.20, 5.5
35.	$P \geqq \frac{1}{2} A \rho^2$	circle	5.9, ρ radius of max. inscribed circle

Cylindrical Condensers

36.	$\frac{L}{2(L'-L)} < c \leqq \frac{1}{2 \log \frac{L'}{L}}$		coatings parallel (24)
37.	$\frac{1}{2 \log \frac{L'}{L}} \leqq c \leqq \frac{B_a}{4\pi \log \frac{L'}{L}}$		1.20, 3.5; coatings similar and similarly situated

1.22. An application. In order to illustrate the use of our table we wish to estimate the principal frequency of a rhombic membrane.

We are given the shorter diagonal d and the acute angle δ of the rhombus. We are required to include Λ between bounds which are conveniently computable.

We begin by surveying the information about Λ available in our table. We find three immediately useful entries, 13, 15 and 33, which give the bounds

(1) $$\Lambda < j\dot{r}^{-1} \ ,$$

(2) $$\Lambda > j\pi^{1/2} A^{-1/2} \ ,$$

(3) $$\Lambda^2 < j^2 B (2A)^{-1} \ .$$

Looking around for other possibilities, we may observe that the rhombus, symmetrized with respect to a line perpendicular to one of its sides, goes over into a rectangle. By our theorem on symmetrization (1.10), the principal frequency of this rectangle is lower than that of the rhombus. On the other hand, the principal frequency of a rectangle is well known, and so we obtain, after easy computation, another lower bound:

(4) $$\Lambda > \pi d^{-1}(1 + \sin^2 \delta)^{1/2} \Big/ \cos (\delta/2) \ .$$

We may also observe that there is another rectangle that has the opposite relation to the rhombus. This rectangle goes over into the rhombus by symmetrization; one of the diagonals of this rectangle coincides with the longer diagonal of the rhombus, the shorter diagonal of which is the line of symmetrization. By our theorem on symmetrization, the principal frequency of this rectangle is higher than that of the rhombus for which we obtain so another upper bound

(5) $$\Lambda < 2\pi d^{-1} \ .$$

To sum up, we have three upper bounds, (1), (3) and (5), and two lower bounds, (2) and (4). It remains to pick out the best bound of each kind. Such picking out needs, of course, more detailed work, although, just in the present case, we can replace various details of calculation by more intuitive considerations.

We can compare intuitively the lower bounds (2) and (4). In fact, (4) was obtained by symmetrizing the given rhombus into a rectangle - a process which diminishes the principal frequency. Symmetrizing this rectangle repeatedly we arrive ultimately at a circle with equal area, the principal frequency of which yields the bound (2). This bound (2), obtained by additional symmetrization, must be lower than (4). Therefore, we reject (2) in favor of (4) which we retain.

A short computation shows that (3) can be written in the form

(3') $$\Lambda < j\ \rho^{-1}$$

where ρ is the radius of the circle inscribed in the rhombus. Let us call r_0 the inner radius of the rhombus with respect to its center. Then certainly

(6) $$\rho < r_0 \leqq \dot{r} .$$

It follows from this and from (1) that

(7) $$\Lambda < j\ r_0^{-1} .$$

On the other hand, it follows from (6) that the upper bound given by (7) is closer than that given by (3'). Therefore, we reject (3'), that is (3), and retain (7). Dodging for the moment the problem of deciding, whether r_0, the inner radius with respect to the center, is or is not the maximum inner radius $\dot{r}$, we abandon also (1). We retain so, eventually, two upper bounds, (5) and (7).

Now, r_0 can be actually computed. We omit the details and give here only an elegant form of the result, obtained by introducing α, the side of the rhombus:

(8) $$r_0 = 4\pi^{1/2} \alpha / [\Gamma(\delta/2\pi)\ \Gamma((\pi - \delta)/2\pi)] .$$

By substituting this value of r_0 in (7) and using a formula on the Γ function, we finally obtain

(9) $$\Lambda < d^{-1}\ j\,\pi^{1/2}\ \Gamma((\pi - \delta)/2\pi)/[2\,\Gamma((2\pi - \delta)/2\pi)] .$$

Recapitulating, we have two upper bounds, (5) and (9) and the lower bound (4). For $\delta = 90^\circ$, that is, in the extreme case of the square, (4) and (5) give the correct value and (9) gives a value slightly higher. Yet, as numerical computation shows, for $\delta = 89^\circ$, and for all angles under 89°, (9) is preferable to (5). Using (4) and (9), we can determine the principal frequency of the rhombus with an error less than 1% when δ is between 75° and 90°. The following table (from 21) exhibits numerical values of the bounds for $\Lambda d/\pi$, given by (4) and (9). The first column expresses δ in degrees, yet in the heading of the last column δ must be measured in radians.*

---

* If from two given numerical values, one of which is a lower bound and the other an upper bound for a certain quantity, we wish to derive an approximation subject to the minimum relative error, we should not take the arithmetic mean of these bounds, but their harmonic mean. See G. Pólya, On the harmonic mean of two numbers, American Math. Monthly, vol. 57 (1950) pp. 26-28.

δ	$\dfrac{(1 + \sin^2 \delta)^{1/2}}{\cos(\delta/2)}$	$\dfrac{j\,\Gamma[(\pi - \delta)/2\pi]}{2\pi^{1/2}\,\Gamma[(2\pi - \delta)/2\pi]}$
90°	2.0000	2.0071
89	1.9826	1.9898
88	1.9654	1.9728
87	1.9483	1.9562
86	1.9313	1.9400
85	1.9145	1.9241
80	1.8322	1.8491
75	1.7525	1.7810
70	1.6752	1.7190
65	1.6002	1.6621
60	1.5273	1.6100
45	1.3256	1.4758
30	1.1575	1.3674
15	1.0419	1.2778
0	1.0000	1.2024

1.23. Comments. Disregarding the detail of the foregoing reasoning, we may retain its general trend: the table in 1.21 and the general theorem on symmetrization are capable of yielding acceptable numerical bounds in concrete cases.

We may also return to a particular point of the foregoing discussion. It is worth observing that not only r_0, the inner radius of the rhombus with respect to its center, can be computed, but also $\bar{r}$, its outer radius, and that, moreover,

$$\pi r_0 \bar{r} = A . \tag{1}$$

We can express the contents of (1) by saying that the area-radius is the geometric mean between the outer radius and the inner radius with respect to the center. By the way, exactly the same relation holds between these three radii also in the case of any regular polygon.

The concrete case treated in 1.22 is rich in suggestive detail, as concrete cases often are. For example, it suggests the following theorem:

> If a convex domain has a center of symmetry, the maximum $\dot{r}$ of the inner radius r_a is attained when a coincides with the center of symmetry.

(The point c is called here a "center of symmetry" if the domain admits a rotation of $2\pi/m$ about c where m is an integer, $m \geq 2$.) It is easy to show by examples that this statement becomes false if the restriction "convex" is dropped. A special case can be proved by symmetrization; see 7.5 (a). [The full statement can be proved by the theory of conformal mapping;

see 7.5 (b) and 9, where also relation (1), in the form

(2) $$\pi \dot{r} \bar{r} = A ,$$

is extended to an arbitrary triangle.]

Methods Based on a Minimum Property

1.24. Analogy. The following sections explain the method by which the majority of our inequalities for C, P and Λ have been obtained. This method is based on certain analogous properties of these three physical quantities. In the three next sections, we shall survey the most essential properties of this kind but we can hint here briefly the physical reason for the analogy that our survey will reveal: The potential energy can be expressed in all three cases as a volume (or area) integral of the square of the gradient. Yet the potential energy must be a minimum under the conditions of the physical problem, and so we are led to questions of the calculus of variations and to differential equations which are different in the different cases but closely analogous to each other.

1.25. Capacity. We consider functions $f(x,y,z)$ defined in the infinite space outside a given solid body B and on the surface A of B which satisfy two boundary conditions:

(1) $$\begin{aligned} f &= 0 \quad \text{at infinite distance, and} \\ f &= u_0 \quad \text{on } A ; \end{aligned}$$

u_0 denotes a given positive constant. The function f is arbitrary otherwise (that is, satisfies only conditions of "smoothness" of a general nature). Then, necessarily,

(2) $$\frac{\iiint (f_x^2 + f_y^2 + f_z^2)\, dx\, dy\, dz}{4\pi u_0^2} \geq C .$$

The triple integral is extended over the whole space outside B; of course, $f_x = \partial f/\partial x$, etc.

Equality in (2) is attained if, and only if, $f(x,y,z) = u(x,y,z)$; the function u is characterized by the following conditions:

(3) $$u_{xx} + u_{yy} + u_{zz} = 0$$

throughout the whole space outside B,

$u = 0$ at infinite distance, and

$u = u_0$ on A.

That is, u is the particular function f (subject to (1)) which satisfies Laplace's equation. In fact, u is the electrostatic potential due to the equilibrium distribution of electricity over the surface A of the conductor B.

What we have stated in this section is essentially a particular case of Dirichlet's principle which will be thoroughly discussed in ch. 2.

1.26. Torsional rigidity. We consider functions f (x,y) defined in the interior of a given plane domain D and on the boundary C of D which satisfy the boundary condition

(1) $$f = 0 \quad \text{on } C .$$

Then, necessarily,

(2) $$\frac{\iint (f_x^2 + f_y^2)\, dx\, dy}{4\left(\iint f\, dx\, dy\right)^2} \geq \frac{1}{P} ;$$

both integrals are extended over the domain D.

Equality in (2) is attained if, and only if,

$$f = cv$$

where c is a constant different from zero and the function v is characterized by the following conditions:

(3) $$v_{xx} + v_{yy} + 2 = 0$$

throughout the interior of D, and

$$v = 0$$

on the boundary C of D. In fact, v is the stress function, the partial derivatives of which determine the components of stress.

For a proof, see 5.2.

1.27. Principal frequency. The function f(x,y) satisfies the same conditions as in the preceding sect. 1.26: it is defined in the interior and on the boundary of D,

(1) $$f = 0 \quad \text{on } C ,$$

and f is arbitrary otherwise. Then, necessarily,

(2) $$\frac{\iint (f_x^2 + f_y^2)\, dx\, dy}{\iint f^2\, dx\, dy} \geq \Lambda^2 ;$$

both integrals are extended over the domain D.

Equality in (2) is attained if, and only if,

$$f = cw$$

where c is a constant different from zero and the function w is characterized by the following conditions:

(3) $$w_{xx} + w_{yy} + \Lambda^2 w = 0$$

and

(4) $$w > 0$$

throughout the interior of D, and

$$w = 0$$

on the boundary C of D. In fact, w characterizes the shape of the membrane when it vibrates emitting its gravest tone.

Compare 5.3.

1.28. Twofold characterization. The parallel facts presented in the foregoing sections show a remarkable analogy between C, P^{-1} and Λ^2. Each of these quantities can be characterized in two ways: either as the value of the minimum in a certain variational problem, or as a certain constant connected with a boundary value problem concerning a certain partial differential equation. In 1.25 (2), 1.26 (2) and 1.27 (2), C, P^{-1} and Λ^2 appear, respectively, as the value of a minimum. If we succeed in solving the differential equations 1.25 (3), 1.26 (3) and 1.27 (3) under the appropriate boundary conditions, we can express C, P^{-1} and Λ^2 from 1.25 (2), 1.26 (2) and 1.27 (2), respectively: these inequalities go over into equations if the proper solution of the corresponding differential equation is substituted for f.

Thus, in order to obtain the exact value of C, P or Λ, we have to solve a boundary value problem. In order to obtain an upper bound for C, P^{-1} or Λ^2, it is enough, however, to substitute for f, in 1.25 (2), 1.26 (2) or 1.27 (2), a function which is almost arbitrary, since it has to satisfy only a simple boundary condition, 1.25 (1), 1.26 (1) or 1.27 (1).

How should we choose this function in order to find a reasonably simple and close bound? This is our next question.

1.29. The choice of level surfaces. We shall be in a better position to understand the procedure of the present book if we recall first the well known method of Lord Rayleigh and Ritz. This method consists in substituting a modified problem for the proposed problem: an elementary extremum problem for the original variational problem. The modification consists essentially in restricting our choice. In both problems, original and substitute, we seek a function minimizing (say) a certain integral. The difference is that originally we have to pick out such a function from a vast class and in the modified problem from a restricted subclass. This subclass of functions depends linearly on a finite number of parameters, and so the modified problem is more elementary than the proposed one. Yet we obtain this advantage at a price: the modified problem gives only an upper bound for the quantity, and not its exact value at which we originally aimed.

Our method applies to the same physical problems, listed in the foregoing sections, to which the Rayleigh-Ritz method is most frequently applied. Also our method consists in substituting for a variational problem, in which the minimizing function depends on two or three independent variables, a more elementary problem: a variational problem with only one independent variable. For example, if we wish to find the capacity C of a solid, we have to minimize the integral on the left hand side of 1.25 (2). The function f depends on three independent variables, the coordinates x, y and z. We modify this problem by restricting our choice to such functions f which have <u>given level surfaces</u>. These level surfaces form a one-parameter family. Therefore, after the selection of the level surfaces, f becomes a function of one variable only and our variational problem is greatly simplified. Of course, we have to pay a price for this simplification: we do not obtain the exact value of the capacity C (at least not in general), only an upper bound for it.

As it is well known, practical success in the applications of the Rayleigh-Ritz method depends mainly on a suitable preliminary choice, and the same is true of our method. We have to choose the level surfaces. In this choice we may be guided by heuristic motives, as some intuitive idea of the true shape of the level surfaces, or the simplicity of formulas. We may, however, be guided also by more definite considerations. We have to define the level surfaces purely geometrically, and this definition should be applicable to all solids (or to a comprehensive subclass of solids, as the convex solids). In general, these geometrically defined surfaces should "look" like the true level surfaces, but in some important particular case (in the case of the sphere, for instance) they should actually coincide with the known level surfaces. Then we know that our method will give the exact result in that particular case (for the sphere) and we may reasonably hope that it may give a fairly good approximation in neighboring cases (for almost spherical solids).

In the foregoing, we paid special attention to level surfaces in space and to the computation of the capacity C which will be considered in detail in ch. 3. Our considerations apply, however, with obvious changes also to level lines in a plane and to the computation of P and Λ, which will be considered in ch. 5.

1.30. <u>The principles of Dirichlet and Thomson</u>. The estimates just discussed are one-sided; they give upper bounds for C, P^{-1} and Λ^2. For C, however, we can obtain also lower bounds by a closely analogous procedure. We shall now discuss both kinds of bounds together, repeating some of the things we have said before in a new context.

We are given a solid $\underline{B}$. We call $\underline{A}$ the surface of $\underline{B}$. We consider $\underline{B}$ as a conductor and examine the electrostatic field outside A. This field is an irrotational and sourceless vector field.

Since the electrostatic field is irrotational, it has a potential. This potential has a constant value u_0 on A and is 0 at infinite distance.

Since the electrostatic field is sourceless, the flux vanishes across any closed surface the interior of which is entirely inside the field. The flux across A does not vanish, however, but is equal to $4\pi Q$, where Q denotes the charge on A.

The energy of the field is the volume integral of the square of the absolute value of the vector defining the field (the electrostatic intensity) extended over the whole field and multiplied by a constant ($1/8\pi$ in vacuum).

We consider now an arbitrary vector field defined outside A. We call "energy" of this field the volume integral of the square of the absolute value of the defining vector extended over the whole space outside A and multiplied by the above constant ($1/8\pi$). The electrostatic field is distinguished from all other fields of this kind by two minimum properties.

> I. Of all irrotational vector fields defined outside A the potential of which has the given values u_0 on A and 0 at infinite distance, the electrostatic field has the minimum energy.

> II. Of all sourceless vector fields defined outside A the flux of which across A has the given value $4\pi Q$, the electrostatic field has the minimum energy.

By theorem I, the irrotational field minimizing the energy must be also sourceless. By theorem II, the sourceless field minimizing the energy must be also irrotational. In statement I the constant u_0 is given; the field with minimum energy must have the flux $4\pi Q$ across A. In statement II the charge Q is given; the field with minimum energy must have the difference of potential u_0 between A and an infinitely distant sphere. In brief, if we assume one half of the properties of the electrostatic field and take as given one half of its characteristic data, then the remaining half of properties and data result from the postulate of minimum energy.

Principle I yields an upper estimate for the capacity C of B in terms of an almost arbitrary irrotational vector field. This estimate is expressed by 1.25 (1) and 1.25 (2). Principle II yields a lower estimate for C in terms of an almost arbitrary sourceless vector field. We do not write here the formula which will be thoroughly considered later (ch. 2). We have discussed at length (in 1.29) how we can use the estimate based on I (the inequality 1.25 (2)) in deriving upper bounds for C. We can foresee now by analogy how we can use II in deriving lower bounds for C. In the first case, it was advantageous to modify the problem by choosing the level surfaces. We can foresee that in the second case it will be advantageous to choose the lines of force.

Theorems I and II are particular cases of more general statements, usually called the principles of Dirichlet and Thomson, respectively. An important special case of Thomson's principle however, was already known to Gauss; referring to this case, we shall speak of the principle of Gauss. That the two principles can be used to obtain estimates of the capacity from two opposite sides, has been observed by Maxwell to whom the method was suggested by an investigation of Lord Rayleigh (15, vol. 1, pp. 138-142, and 26, vol. 2, pp. 175-189). Maxwell, however, did not derive upper or lower bounds from his method in concrete cases, observing that the "operations... are in general too difficult for practical purposes."

Expansions and Variational Methods

1.31. Expansions. Let C denote as before a closed plane curve. There are various methods of characterizing C analytically by the expansion of some connected function. We list a few cases.

1. Parametric representation in cartesian coordinates $x(t)$, $y(t)$. If the range of the parameter t is the interval $-\pi < t \leq \pi$,

$$z(t) = x(t) + i\, y(t)$$

is a periodic function of t with the Fourier expansion

$$z(t) = \sum_{n=-\infty}^{\infty} z_n e^{int} .$$

2. Representation in polar coordinates r, φ. If the curve is star-shaped with respect to the origin and its equation is $r = f(\varphi)$, the function $f(\varphi)$ has the period 2π and the expansion

$$f(\varphi) = \sum_{n=-\infty}^{\infty} f_n e^{in\varphi} .$$

3. Representation in tangential coordinates p, θ. The curve C is convex and contains the origin. Let $p = g(\theta)$ be the distance from the origin to the tangent of C the exterior normal to which includes the angle θ with a fixed direction. Then $g(\theta)$ has the period 2π and the expansion

$$g(\theta) = \sum_{n=-\infty}^{\infty} g_n e^{in\theta} .$$

4. Conformal representation of the interior of C. We let C lie in the complex z plane and map the interior of C onto the circle $|\zeta| < 1$. The mapping function can be expanded in a power series,

$$z = c_0 + c_1 \zeta + c_2 \zeta^2 + c_3 \zeta^3 + \dots$$

and is uniquely determined if c_0, the image of $\zeta = 0$, is given and c_1 is assumed to be positive.

5.' Conformal mapping of the exterior. We map the exterior of C onto $|\zeta| > 1$ so that $z = \infty$ and $\zeta = \infty$ correspond to each other. Then

$$z = d\zeta + d_0 + d_1 \zeta^{-1} + d_2 \zeta^{-2} + \dots .$$

We assume that d is positive.

We have listed five different sequences of coefficients, z_n, f_n, g_n, c_n, and d_n, connected with the curve C. The area A of C can be elegantly expressed in terms of all five sequences. Not only A, but also P, the torsional rigidity, can be expressed in terms of the coefficients connected with the conformal mapping of the interior, and we shall use these expressions in establishing an important inequality between A and P; see 5.17. We shall use the expression of the exterior mapping function in proving an inequality between $\bar{r}$ and C, the capacity of a solid of revolution and another inequality between $\bar{r}$ and I; see 3.7 and 5.19, respectively.

Hurwitz considered tangential coordinates in space and the corresponding expansion in spherical harmonics, and succeeded in expressing M, S and V in terms of this expansion*.

1.32. Nearly circular curves and nearly spherical surfaces. The geometrical and physical quantities, depending on a curve or a surface, which we are investigating have not yet all been expressed in terms of the coefficients of a suitable expansion. Yet we shall express so all of them, with one exception, restricting ourselves to the case of nearly circular curves and nearly spherical surfaces. In order to explain more clearly the nature of the expressions which we have in mind we need an example.

Lord Rayleigh encountered the first example of this kind in investigating the principal frequency of a nearly circular membrane (26, vol. 1, pp. 339-342). Let

$$r = 1 + \rho(\varphi) = 1 + \delta\bar{\rho}(\varphi) \tag{1}$$

be the equation of the boundary of the membrane in polar coordinates r and φ. The periodic function $\bar{\rho}(\varphi)$ is fixed, δ is a variable restricted to a (small) neighborhood of 0, and $\rho(\varphi) = \delta\bar{\rho}(\varphi)$ represents the infinitesimal

* Sur quelques applications géométriques des séries de Fourier. Annales de l'École Normale Supérieure, series 3, vol. 19 (1902) pp. 357-408; Mathematische Werke, vol. 1, pp. 509-554.

variation of the unit circle. We consider the Fourier series

$$(2) \qquad \rho(\varphi) = a_0 + 2\sum_{n=1}^{\infty} (a_n \cos n\varphi + b_n \sin n\varphi) .$$

Each coefficient a_n or b_n is of the first order, or order δ; that is, it represents a fixed number multiplied by the infinitesimal variable δ. Rayleigh expanded $\bar{\Lambda}$, the radius of the circle with principal frequency Λ, in powers of δ, and, neglecting terms of higher than second order, found that

$$(3) \qquad \bar{\Lambda} = 1 + a_0 - \sum_{n=1}^{\infty} \left(1 + \frac{2j\, J_n'(j)}{J_n(j)} \right) (a_n^2 + b_n^2) .$$

The abbreviation j has been explained above; see 1.3 (1). The precise meaning of the equation (3) is that the difference of the two sides, divided by δ^2, tends to 0 when $\delta \to 0$. The value of $\bar{\Lambda}$ when $\delta = 0$, that is 1, is the first term on the right hand side of (3). The next term a_0, which is proportional to δ, is the first variation of $\bar{\Lambda}$, and the last term (the infinite sum with reversed sign) proportional to δ^2, is the second variation. Rayleigh used formula (3) in an attempt to prove his conjecture about the principal frequency of the circular membrane which was later fully established by Faber and Krahn and connected with symmetrization in our previous work.

In brief, Rayleigh considers a quantity (in his example, $\bar{\Lambda}$), depending on the form of the boundary. The distinctive feature of Rayleigh's method is to expand the variation of the circular boundary in a Fourier series and to express the first and second variations of the quantity dependent on the boundary in terms of the Fourier coefficients. We applied this method systematically and extended it by analogy from plane to space, passing from Fourier expansions to expansions in spherical harmonics. We survey our results in the next section.

1.33. Variational formulas in polar coordinates. Table. We begin by considering the quantities of which we shall compute the variation. They are listed in the table at the end of this section under the heading Q, beginning with $\bar{L}$, $\bar{r}$, ... Most of these quantities have been defined in 1.4 and 1.5 but a few need special comment.

$\bar{I}_c = \bar{I}$, as defined in 1.3 and 1.5; c stands here for the centroid of of the area bounded by C, which is conceived as covered with matter with surface density 1.

r_c in no. 7 has the usual meaning (1.3): it denotes the inner radius of the curve C with respect to the point c just defined.

r_c in no. 16 has a different but analogous meaning: it denotes the inner radius of the closed surface A (defined in 32) with respect to the centroid c of the surrounded volume, which we conceive as a material solid of uniform density.

[Λ_1, Λ_2, and Λ_3 arise from three parallel problems on elastic deformation: $\Lambda_1 = \Lambda$ (in the notation used from 1.3 till 1.32) is the principal frequency of a membrane with boundary C, Λ_2 the principal frequency of a clamped plate with the same boundary, and Λ_3 is connected with the buckling of a plate of this shape. Each of the quantities Λ_1, Λ_2 and Λ_3 is the first eigen-value of a partial differential equation, namely

$$(1) \qquad \nabla^2 u + \Lambda_1^2 u = 0, \qquad \nabla^4 u - \Lambda_2^4 u = 0, \qquad \nabla^4 u + \Lambda_3^2 \nabla^2 u = 0,$$

respectively. The function u satisfies the boundary condition

$$(2') \qquad u = 0 \quad \text{on } C$$

in the first case,

$$(2'') \qquad u = \frac{\partial u}{\partial n} = 0 \quad \text{on } C$$

in the second and third case. If C is the unit circle,

$$\Lambda_1 = j, \qquad \Lambda_2 = h, \qquad \Lambda_3 = k$$

where j, h and k denote the minimum positive root of the Bessel functions

$$J_0(x), \qquad J_0(x)\, I_0'(x) - J_0'(x)\, I_0(x), \qquad J_0'(x),$$

respectively. Numerically

$$j = 2.4048, \qquad h = 3.1962, \qquad k = 3.8317 .$$

In accordance with the convention laid down in 1.5, $\bar{\Lambda}_1$, $\bar{\Lambda}_2$ and $\bar{\Lambda}_3$ denote the radii of three circles which agree with the curve C in the value of Λ_1, Λ_2 and Λ_3, respectively. The full definition of these quantities will be discussed in note F.

W_m and P_m stand for average virtual mass and average polarization, respectively, and $\bar{W}_m$ and $\bar{P}_m$ for the corresponding radii. For the full definition see note G and for the derivation of the variational formulas 33 and 28.]

The first part of the table, nos. 1-9, is concerned with a nearly circular curve C, characterized by equations 1.32 (1) and 1.32 (2) and lists expansions of the form

$$(3) \qquad Q = 1 + a_0 + \sum_{n=1}^{\infty} R(n)\,(a_n^2 + b_n^2) .$$

The values for Q and R(n) must be taken, of course, from the same horizontal line of the table and from the respective columns. For instance, no. 6 lists, as a particular case of (3), Rayleigh's formula 1.32 (3) for $\Lambda = \bar{\Lambda}_1$. The precise meaning of (3) is

$$\lim_{\delta \to 0} \{Q - a_0 - \sum_{n=1}^{\infty} R(n)\,(a_n^2 + b_n^2)\}\,\delta^{-2} = 0 \,. \tag{4}$$

Therefore a_0, of order δ, represents the first variation of Q, and the infinite sum, of order δ^2, the second variation. The expression for R(n), given in the table, is valid for n = 2, 3, 4,..., but not necessarily for n = 1. We lay down the <u>convention</u> that

$$R(1) = 1 \,. \tag{5}$$

This convention is necessary for nos. 3 and 7, marked by * in the table; in the other cases, the given formula yields (5) automatically. Thus, no. 3 and no. 7 mean the expansions

$$\bar{I}_c = 1 + a_0 + a_1^2 + b_1^2 + 3(a_2^2 + b_2^2) + 3(a_3^2 + b_3^2) + \ldots \,, \tag{6}$$

$$r_c = 1 + a_0 + a_1^2 + b_1^2 - 5(a_2^2 + b_2^2) - 7(a_3^2 + b_3^2) - \ldots \,. \tag{7}$$

The second part of the table, nos. 10 - 16, is concerned with a nearly spherical surface <u>A</u>, the equation of which in spherical coordinates r, θ, φ, is

$$r = 1 + \rho(\theta, \varphi) = 1 + \delta\,\bar{\rho}(\theta, \varphi) \,. \tag{8}$$

Here $\bar{\rho}(\theta, \varphi)$ denotes a fixed function defined in the points of the unit sphere, δ a variable restricted to a neighborhood of 0, and $\rho(\theta, \varphi) = \delta\,\bar{\rho}(\theta, \varphi)$ the variation of the unit sphere. We expand $\rho(\theta, \varphi)$ in a Laplace series

$$\rho = \sum_{n=0}^{\infty} X_n(\theta, \varphi) \,. \tag{9}$$

The term $X_n(\theta, \varphi)$ represents a spherical (surface) harmonic of degree n. All terms X_n are of order δ. That is, $X_n = \delta \bar{X}_n$, where $\bar{X}_n$ is a fixed spherical harmonic.

The second part of the table lists expressions of the form

$$Q = 1 + X_0 + \sum_{n=1}^{\infty} R(n)\,\frac{1}{4\pi} \iint [X_n(\theta, \varphi)]^2\, d\omega \,. \tag{10}$$

The double integral is extended over the surface of the unit sphere of which $d\omega$ is an element. The precise meaning of (10) is similar to that of (3), which is explicitly stated by (4). For instance, no. 14 reads

(11) $$C = 1 + X_0 + \frac{1}{4\pi}\iint X_1^2\,d\omega + \frac{2}{4\pi}\iint X_2^2\,d\omega + \frac{3}{4\pi}\iint X_3^2\,d\omega + \dots .$$

The convention (5) holds. It applies to no. 12 and no. 16.

Variational formulas of the form (10) have been first given, as far as we know, in a paper by Szegö (33).

[The nos. 9 and 12 have been added since September 1948. The corresponding small additions in 1.35 (b) are, exceptionally, not marked with square brackets.]

Nearly circular curves

No.	Q	R(n)
1	$\bar{L}$	n^2
2	$\bar{r}$	$2n - 1$
3*	$\bar{I}_c$	3
4	$\bar{A}$	1
5	$\bar{P}$	$-2n + 3$
6	$\bar{\Lambda} = \bar{\Lambda}_1$	$-1 - \frac{2j\,J_n'(j)}{J_n(j)}$
7*	r_c	$-2n - 1$
8	$\bar{\Lambda}_2$	$1 + \frac{2hJ_0'(h)}{-J_0(h)} - 4h\left(\frac{I_n'(h)}{I_n(h)} - \frac{J_n'(h)}{J_n(h)}\right)^{-1}$
9	$\bar{\Lambda}_3$	$1 - 2k\left(\frac{n}{k} - \frac{J_n'(k)}{J_n(k)}\right)^{-1}$

Nearly spherical surfaces

No.	Q	R(n)
10	$\bar{M}$	$\frac{n^2 + n}{2}$
11	$\bar{S}$	$\frac{n^2 + n + 2}{4}$
12*	$\bar{P}_m$	$1 + \frac{3n(2n - 1)}{2(2n + 1)}$
13	$\bar{W}_m$	$1 + \frac{3(n-1)(2n-1)}{2(2n + 1)}$

Nearly spherical surfaces (cont.)

No.	Q	R(n)
14	C	n
15	$\bar{V}$	1
16*	r_c	-n -1

1.34. Variational formulas in tangential coordinates. The equation of a nearly circular curve in tangential coordinates p, θ is of the form

(1) $$p = 1 + h(\theta) = 1 + \delta \bar{h}(\theta) ;$$

$\bar{h}(\theta)$ is a fixed periodic function. We expand the variation $h(\theta)$ in a Fourier series

(2) $$h(\theta) = h_0 + 2 \sum_{n=1}^{\infty} (h_n \cos n\theta + k_n \sin n\theta) .$$

All the quantities Q listed in the first part of the foregoing table can be expressed also in terms of the coefficients of the expansion (2). In fact, if R(n) has the same meaning as in the expansion 1.33 (3) or in the table,

(3) $$Q = 1 + h_0 + \sum_{n=2}^{\infty} (R(n) - n^2)(h_n^2 + k_n^2) .$$

The precise interpretation of this equation is similar to that of 1.33 (3) given by 1.33 (4). For example, no. 1 of the table yields the extremely simple result

(4) $$\bar{L} = 1 + h_0 .$$

The equation of a nearly spherical surface in tangential coordinates p, θ', φ' is of the form

(5) $$p = 1 + h(\theta', \varphi') = 1 + \delta \bar{h}(\theta', \varphi') ;$$

$\bar{h}(\theta', \varphi')$ is fixed. We expand $h(\theta', \varphi')$ in a Laplace series

(6) $$h(\theta', \varphi') = \sum_{n=0}^{\infty} Y_n(\theta', \varphi') .$$

Then, if R(n) has the same meaning as in 1.33 (10) or in the table, we have for any quantity Q listed there under nos. 10 - 16

(7) $$Q = 1 + Y_0 + \sum_{n=2}^{\infty} \left(R(n) - \frac{n^2 + n}{2}\right) \frac{1}{4\pi} \iint [Y_n(\theta', \varphi')]^2 \, d\omega' .$$

For example, no. 10 of the table yields the extremely simple result

$$\bar{M} = 1 + Y_0 . \tag{8}$$

The variational formulas (3) and (7) are not only remarkably elegant but also useful in computing and checking the table. In particular, the connection between (7) and 1.33 (10) allows us to profit from Hurwitz's results mentioned toward the end of 1.31.

1.35. <u>A necessary condition</u>. (a) If Q' and R'(n) are so connected as Q and R(n) are in 1.33 (3), we obtain, in view of 1.33 (4), that

$$\lim_{\delta \to 0} \left\{ Q - Q' - \sum_{n=2}^{\infty} [R(n) - R'(n)] (a_n^2 + b_n^2) \right\} \delta^{-2} = 0 . \tag{1}$$

We notice here the quadratic form of an infinity of variables

$$\sum_{n=2}^{\infty} [R(n) - R'(n)] (a_n^2 + b_n^2) \tag{2}$$

to which we can apply with advantage the usual terminology. If

$$R(n) - R'(n) > 0 \qquad \text{for } n = 2, 3, \ldots$$

we say that the form (2) is positive definite. If

$$R(n) - R'(n) \geq 0 \qquad \text{for } n = 2, 3, \ldots$$

and equality is attained for at least one value of n, we say that (2) is positive semi-definite. If among the differences R(n) - R'(n) at least one is positive and at least one negative, we call (2) indefinite. With this terminology, we can prove, as in the finite case, the following lemma:

> A necessary condition for the validity of $Q \geq Q'$ for all nearly circular curves is that the form (2) should be positive, definite or semi-definite.

In fact, if (2) were neither positive definite nor positive semi-definite, there would exist an m such that

$$R(m) - R'(m) < 0, \qquad m \geq 2 .$$

Yet then we could take

$$a_n^2 + b_n^2 = \begin{cases} 0 & \text{for } n \neq m , \\ \delta^2 & \text{for } n = m \end{cases}$$

and (1) would show that $Q < Q'$ for sufficiently small δ .

Of course, there is a formula corresponding to (1) and dealing with expansions of the form 1.33 (10), to which the terminology introduced and the lemma proved apply without essential change. We shall find this useful in surveying the facts condensed in our table 1.33.

(b) Let Q and Q' denote two different quantities which belong to the same part of the table (to the first part, if both deal with curves, and to the second, if both deal with surfaces). Let Q precede Q'. The corresponding quadratic form is positive definite in all but five cases. The five exceptional pairs are

$$r_c\ ,\ \bar{\Lambda}_2\ ;\qquad \bar{S}\ ,\ \bar{P}_m\ ;\qquad \bar{W}_m\ ,\ C\ ;\qquad \bar{r}\ ,\ \bar{I}_c\ ;\qquad \bar{S}\ ,\ C\ .$$

The corresponding form is indefinite in the first three cases, and positive semi-definite in the last two. This is immediately verified in all cases not involving nos. 6, 8 or 9, for which see 6.5, 6.6 and 6.6A. We consider now the exceptional cases.

The form corresponding to r_c and $\bar{\Lambda}_2$ is indefinite. By our lemma, there are some nearly circular curves for which $r_c > \bar{\Lambda}_2$ and others for which $r_c < \bar{\Lambda}_2$. In short, r_c and $\bar{\Lambda}_2$ are not comparable and the same is true of $\bar{S}$ and $\bar{P}_m$, and also of $\bar{W}_m$ and C.

The form corresponding to $\bar{r}$ and $\bar{I}_c$ is semi-definite. (The difference $R(n) - R'(n)$ vanishes for $n=2$ but is positive for $n \geq 3$.) It can be shown by other means that $\bar{r} \geq \bar{I}_c$ for any curve; see 5.19.

The form corresponding to $\bar{S}$ and C is also semi-definite. (The difference $R(n) - R'(n)$ behaves exactly as in the foregoing case.) Yet, $\bar{S}$ and C are not comparable; for oblate spheroids $\bar{S} > C$, for prolate spheroids $\bar{S} < C$ (24), and such spheroids can differ from a sphere as little as we please.

(c) It is interesting to look into the details of the last case. The equation in spherical coordinates r, θ, φ

$$(3)\qquad r^2\,[1 + 2\,\delta P_2\,(\cos\theta)\,] = 1$$

represents an ellipsoid with semiaxes $(1-\delta)^{-1/2}$, $(1-\delta)^{-1/2}$, $(1+2\delta)^{-1/2}$; that is, an oblate or a prolate spheroid according as $\delta > 0$ or $\delta < 0$. For small δ, we can replace (3) by

$$(4)\qquad r = 1 - \delta\,P_2\,(\cos\theta)$$

disregarding terms of the second order. If we start from (4), nos. 11 and 14 of the table give

$$\bar{S} = C$$

correctly to terms of the second order inclusively, since $R(2) = R'(2)$ in this case, and independently of the sign of δ. That the difference $\bar{S} - C$ actually changes sign with δ, is due to terms of the third order which do not appear at all in our variational formulas.

This example shows strikingly that the necessary condition derived at the beginning of this section is by no means sufficient.

(d) Let us now set aside the five cases listed at the beginning of (b). In the remaining cases the form (2) is positive definite. In many of these cases we can prove that $Q \geqq Q'$ is generally true (see 1.13, 1.15) and in no such case could we disprove the conjecture $Q \geqq Q'$.

This observation suggests several interesting questions. We cannot enter, however, into these or into other suggestive details of the table, but we should at least mention one. It is noteworthy that both the first and the second variations of the combinations $\bar{r} - 2\bar{A} + \bar{P}$ and $\bar{M} - 2\bar{S} + \bar{V}$ vanish identically, that is, for all almost circular (or spherical) domains. This suggests approximate formulas, especially the following

$$P \sim A^4 / (2\pi^3 \bar{r}^4)$$

which could have, perhaps, practical applications.

[Another possible practical application is approximate computation. M. Aissen, see 1, computed P for a regular polygon with 20 sides, using the formula no. 5 of the table.]

Further Results and Remarks

1.36. Ellipsoid. The shape of the ellipsoid with semi-axes a, b and c depends only on the ratios a:b:c. In order to visualize the variety of these shapes, we consider a, b and c as homogeneous coordinates of a point (a,b,c) in a plane, and we consider that part of this plane where

$$a \geqq b \geqq c \geqq 0 . \tag{1}$$

These inequalities delimit a triangle which we can make equilateral by choosing suitably the coordinate system; see Fig. 1. The points of this triangle are in one-to-one correspondence with the different ellipsoidal shapes. The interior points correspond to non-degenerate ellipsoids with three different axes, the boundary points to degenerate ellipsoids or to ellipsoids of revolution. The three vertices represent the sphere, the circular disk and the "needle" (segment of a straight line), respectively.

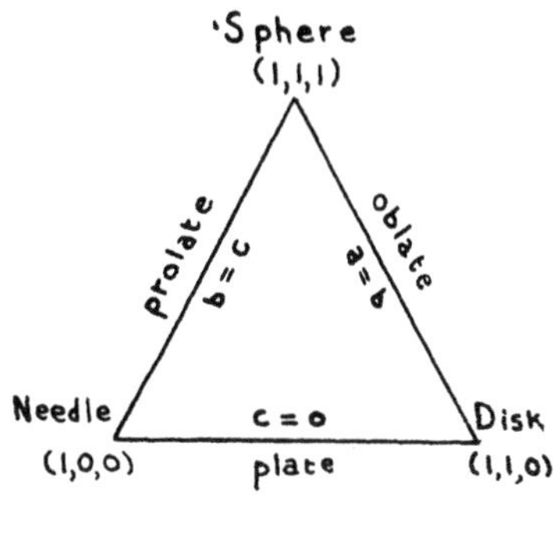

Fig. 1

We can characterize the shape of an ellipsoid in another way, by giving two of the three quantities α, β, and γ, the eccentricities of the three principal sections of the ellipsoid. Under the assumption (1), the non-negative numbers α, β, and γ are defined by

(2) $$1 - \alpha^2 = c^2 b^{-2}, \qquad 1 - \beta^2 = c^2 a^{-2}, \qquad 1 - \gamma^2 = b^2 a^{-2},$$

and linked by the equation

(3) $$1 - \beta^2 = (1 - \alpha^2)(1 - \gamma^2) .$$

Approximating the electrostatic capacity by various geometric quantities is one of our principal aims. Such approximations can be thoroughly studied for ellipsoids, especially for almost spherical ellipsoids, which are represented by a neighborhood of the vertex (1, 1, 1) of the triangle of Fig. 1 where the eccentricities α, β, and γ are small. Let C' be an approximation to C, the capacity of the ellipsoid. The relative error of C' is (C' - C)/ C; we expand it in powers of β and γ. This expansion is a sum of homogeneous polynomials of different degrees; the polynomial of lowest degree that does not vanish identically is called the _initial term of the relative error of the approximation_ C'.

The table at the end of the present section lists eight different approximations to C and the corresponding initial terms. The last column uses the abbreviations

(4) $$[\beta^4] = \beta^4 - \beta^2\gamma^2 + \gamma^4 ,$$

(5) $$[\beta^6] = (\beta^2 + \gamma^2)(2\beta^2 - \gamma^2)(\beta^2 - 2\gamma^2) / 2 .$$

Thus, line no. 1 of the table indicates that

$$(\bar{M} - C) / C = (\beta^4 - \beta^2\gamma^2 + \gamma^4) / 45 + \dots ;$$

the following terms are of 6th, or higher, degree.

Entries 1 - 5 are ordered according to increasing _efficiency in approximating almost spherical ellipsoids_, that is, according to increasing degree of the initial term, and those between which this degree makes no difference, according to decreasing absolute value of a numerical factor. Nos. 1 and 2 are equally effective, but less effective than any of the following, and no. 5 is the most effective linear combination of $\bar{V}$, $\bar{S}$ and $\bar{M}$. Nos. 6 - 8 are also ordered according to increasing efficiency.

In the cases here listed, however, the initial term indicates more than efficiency in the vicinity of the sphere. When the degree of the initial term is 4 (nos. 1, 2, 6 and 7) the error of the approximation keeps a _constant sign for all ellipsoids_. When this degree is 6, the error of the approximation keeps a _constant sign for oblate spheroids, and also for prolate spheroids, but these two signs are opposite_. No. 5, where the degree in question is 8, behaves differently; its error changes sign in the transition from nearly spherical prolate spheroids to the needle. See 8.3 and

8.2 concerning nos. 4 and 8, respectively; nos. 1, 2, 3, 6 and 7 have been discussed previously (24).

The occurrence of the polynomials (4) and (5) in all initial terms is remarkable. This circumstance will be elucidated by a general theorem in sect. 8.1.*)

No.	Approximation	Initial Term
1.	$\bar{M}$	$[\beta^4]/45$
2.	$\bar{V}$	$-[\beta^4]/45$
3.	$\bar{S}$	$2[\beta^6]/945$
4.	$(\bar{M} + \bar{V})/2$	$-4[\beta^6]/2835$
5.	$(3\bar{M} + 4\bar{S} + 3\bar{V})/10$	$-[\beta^4]^2/15750$
6.	$[(bc)^{1/2}+(ca)^{1/2}+(ab)^{1/2}]/3$	$-11[\beta^4]/720$
7.	$[a + b + c]/3$	$[\beta^4]/180$
8.	$\{11[a+b+c]+4[(bc)^{1/2}+(ca)^{1/2}+(ab)^{1/2}]\}/45$	$-11[\beta^6]/15120$

1.37. Lens. (a) If two spheres intersect, each is divided into two zones by the circle of intersection. A lens is a solid bounded by two of these zones belonging to different spheres. Let α and β denote the exterior angles which the two zones bounding the lens include with the plane of the circle of intersection; see Fig. 2. The angles α and β determine the shape of the lens. Without omitting any shape, we can assume that

$$(1) \qquad 0 \leqq \alpha \leqq \beta, \qquad \alpha + \beta \leqq 2\pi .$$

If we regard α and β as rectangular coordinates in a plane the inequalities (1) delimit a right triangle which represents the variety of lenticular shapes; see Fig. 2. The correspondence between shapes and points is one-to-one in so far as non-degenerate lenses are concerned, but this is no longer true in the limiting cases. All the points on the hypotenuse of the triangle, where $\alpha = 0$, and also on another line, parallel to one of the legs, where

*) We use the opportunity to observe that one section of our paper 24 (section 9, which deals with the capacity of the ellipsoid and Landen's transformation) is closely related to the following paper with which we were not acquainted at the time of the publication of 24: R. Hargreaves, An ellipsoidal type of elliptic integrals, Messenger of Math. vol. 36 (1906) pp. 176-188.

$\alpha + \beta = \pi$, represent the same shape, the sphere. On the other hand, the origin represents an infinity of shapes, namely two spheres in contact with an arbitrary ratio of the radii. One of the legs of the triangle in Fig. 2 represents the symmetrical lenses, the other the spherical bowls, and the vertex of the right angle the circular disk. Fig. 2 shows the shapes associated with a few points, some dotted lines along which it is instructive to follow the variation of the shape and, in full lines, a smaller right trangle the points of which represent the convex lenticular shapes.

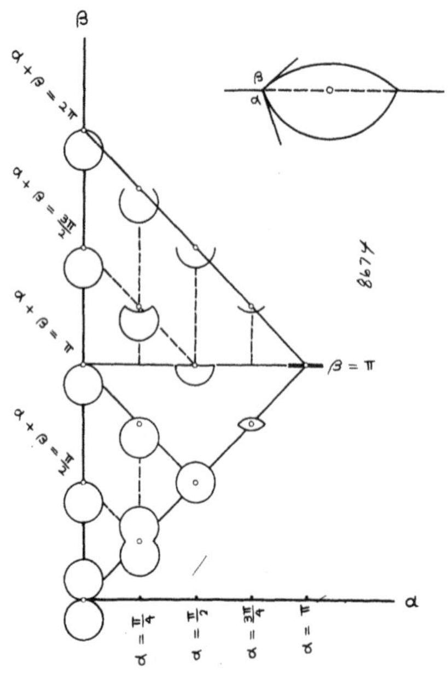

Fig. 2

To discuss the relations between capacity and various geometric quantities is more difficult in the case of the lens than in that of the ellipsoid. J. G. Herriot made an interesting study of the ratios $C/\bar{S}$ and $C/\bar{r}$ (10). Some inequalities concerning these ratios will be proved in 8.4 - 8.6.

(b) Such particular examples as the ellipsoid and the lens could be useful in a continuation of our study which should lead eventually to a complete system of inequalities between the quantities introduced, as C, V, S, M, ... In order to visualize this faraway aim, let the point with rectangular coordinates ξ, η, ζ correspond to a convex solid with volume V, surface-area S, Minkowski constant M and capacity C, by means of the equations

(2) $$\xi = \bar{S}/\bar{M}, \qquad \eta = \bar{V}/\bar{M}, \qquad \zeta = C/\bar{M}.$$

The aim is to determine completely the set X of all such points (ξ, η, ζ). The fundamental inequalities 1.13 (1), 1.13 (2), 1.13 (4) and 1.13 (5), translated into the present notation, are written as follows:

(3) $$0 \leq \xi \leq 1, \qquad 0 \leq \eta \leq \xi^{4/3},$$

(4) $$\eta \leq \zeta \leq (1 - \xi^2)^{1/2} \Big/ \log \{[1 + (1 - \xi^2)^{1/2}]\, \xi^{-1}\}.$$

Thus, the set X occupies a relatively small part of a cube with unit edge. The ellipsoidal and the lenticular shapes yield two surfaces belonging to X. There are three points of which we know with certainty that they belong to the boundary of X. These are the points

$$(1, 1, 1), \quad (0, 0, 0), \quad (8^{1/2}\pi^{-1}, 0, 8\pi^{-2})$$

which correspond to the sphere, the needle and the circular disk, respectively.

1.38. Survey of the following chapters. The foregoing summary of our results and methods is not quite complete. The following brief survey mentions a few additional points.

Ch. 2 prepares for the subsequent applications by discussing briefly The principles of Dirichlet and Thomson and connected topics, especially a general method for computing the capacity of a given solid. Ch. 3 presents Applications of the principles of Dirichlet and Thomson to estimation of the capacity and derives several important inequalities listed above in 1.13 and 1.14. Ch. 4 derives a lower estimate for the capacity of the Circular plate condenser from Gauss' principle. Ch. 5 applies to Torsional rigidity and principal frequency the method of appropriately chosen level lines and other tools, as conformal mapping and an elementary geometrical theorem, the "inclusion lemma." Ch. 6 derives the variational formulas announced above, in 1.33 and 1.34; it considers several important geometrical and physical quantities in the case of Nearly circular and nearly spherical domains. Ch. 7 brings somewhat scattered remarks On symmetrization and ch. 8 On ellipsoid and lens.

[Notes A - F discuss various aspects of symmetrization. Note G deals briefly with virtual mass and polarization.]

CHAPTER II. THE PRINCIPLES OF DIRICHLET AND THOMSON

2.1. Capacity. Energy. In this chapter we discuss in more detail the concepts broadly dealt with in the introduction (1.24, 1.25, 1.30; 15, vol. 1, pp. 138-140).

Let A be a closed surface to which we shall also refer as a "conductor." We define a function u(p) harmonic outside A, assuming the constant value $u(p) = u_0$ on A and having the form

(1) $$u(p) = X_0 r^{-1} + X_1 r^{-2} + X_2 r^{-3} + \dots$$

at infinity; here X_0, X_1, X_2,... denote surface harmonics of degree 0, 1, 2,..., respectively. The ratio $X_0/u_0 = C$ depends only on the conductor A and is called its capacity. It is the charge which, in electrostatic equilibrium on A, raises the constant potential of A to unity. Obviously

(2) $$C = -\frac{1}{4\pi u_0} \iint \frac{\partial u}{\partial n_e}\, d\sigma$$

where the surface integral is extended over an arbitrary closed surface containing A (possibly over A itself); $d\sigma$ is the surface element, n_e the exterior normal.

One of our principal purposes is to find upper and lower bounds for C in terms of certain geometrical quantities associated with A.

Upper bounds for C can be found by means of the principle of Dirichlet.

Lower bounds for C can be found by means of the principle of Thomson.

These assertions are based on the intuitive idea that the energy of an arbitrary charge distribution on or inside of the conducting surface A is always greater than the energy of the distribution which is in electrostatic equilibrium. The energy of the latter distribution can be expressed either as a function of the capacity and potential or as a function of the capacity and the charge. Indeed we have for the energy E the expression

(3) $$E = QV/2$$

where Q is the charge and V the constant potential on A (difference of the potentials on A and at infinity). In the notation used above, $Q = X_0$,

$V = u_0$ so that $Q/V = C$; hence

(4) $$E = CV^2/2 = Q^2/(2C) .$$

2.2. Formulation of the principles. (a) Let $f(p)$ be an arbitrary scalar function defined in the field, that is, outside $\underline{A}$; we assume $f = u_0$ on $\underline{A}$ and $f = 0$ at infinity. The energy of the distribution corresponding to this function is defined by the integral

(1) $$E_1 = (8\pi)^{-1} \iiint |\text{grad } f|^2 \, d\tau$$

extended over the field; $d\tau$ is the volume element. Then $E_1 \geqq E = CV^2/2 = Cu_0^2/2$ so that

(2) $$C \leqq (4\pi)^{-1} u_0^{-2} \iiint |\text{grad } f|^2 \, d\tau$$ (Dirichlet's principle).

(b) Let $\vec{f}(p)$ be an arbitrary sourceless vector function defined in the field, that is,

(3) $$\text{div } \vec{f}(p) = 0 .$$

We assume that

(4) $$(4\pi)^{-1} \iint \vec{f}_n \, d\sigma = Q ,$$

that is, the charge; the domain of integration is the same as above and n stands for n_e; see 2.1 (2). The energy of the distribution corresponding to this function is defined by the integral

(5) $$E_2 = (8\pi)^{-1} \iiint |\vec{f}|^2 \, d\tau$$

extended over the field. Then $E_2 \geqq E = Q^2/(2C)$ so that

(6) $$1/C \leqq (4\pi)^{-1} Q^{-2} \iiint |\vec{f}|^2 \, d\tau$$ (Thomson's principle).

In 2.4 and 2.6 formal proofs for the important inequalities (2) and (6) will be given. In 2.8 we shall point out a special case of (6) due to Gauss. Substituting various special functions $f(p)$ and $\vec{f}(p)$ in the inequalities (2) and (6), satisfying the conditions mentioned above, we obtain various upper and lower bounds for the capacity.

(c) These principles can be formulated, with proper modification, for other types of charge distributions. We mention here two cases the first of which is only slightly different from the case considered above. In both cases we deal with a pair of surfaces to which we refer as a "condenser."

Let $\underline{A}_0$ and $\underline{A}_1$ be two closed surfaces, $\underline{A}_1$ containing $\underline{A}_0$. We define the harmonic function u(p) between $\underline{A}_0$ and $\underline{A}_1$ assuming constant values on $\underline{A}_0$ and $\underline{A}_1$: $u(p) = u_0$ on $\underline{A}_0$ and $u(p) = u_1$ on $\underline{A}_1$, $u_0 \neq u_1$. Then

(7) $$C = \frac{1}{4\pi(u_0 - u_1)} \iint \frac{\partial u}{\partial n_i} d\sigma$$

is called the capacity; this integral is extended over an arbitrary surface containing $\underline{A}_0$ and contained in $\underline{A}_1$; n_i is the interior normal. Then Dirichlet's principle holds in the following form:

(8) $$C \lesssim (4\pi)^{-1} (u_0 - u_1)^{-2} \iiint |\text{grad } f|^2 \, d\tau \;;$$

the integration is extended over the field, that is, the domain between $\underline{A}_0$ and $\underline{A}_1$. The function f satisfies the condition $f = u_0$ on $\underline{A}_0$ and $f = u_1$ on $\underline{A}_1$. Also statement (6) holds without essential change.

(d) The other case is where $\underline{A}_0$ and $\underline{A}_1$ are two closed surfaces lying outside of each other. We define the harmonic function u(p) outside of $\underline{A}_0$ and $\underline{A}_1$, $u(p) = u_0$ on $\underline{A}_0$, $u(p) = u_1$ on $\underline{A}_1$, and without "charge" at infinity. (This means that the expansion 2.1 (1) of u for large r has no term of the form $X_0 r^{-1}$.) In this case we retain the definition (7), only the integration is extended over a surface surrounding $\underline{A}_0$ but leaving outside $\underline{A}_1$. The further changes necessary in this case are obvious.

2.3. Two-dimensional case. The analogy between space and plane is the closest in the case of the problem which is dealt with in 2.2 (c).

Let $\underline{C}_0$ and $\underline{C}_1$ be two closed curves, $\underline{C}_1$ containing $\underline{C}_0$. We define a harmonic function u(p) between $\underline{C}_0$ and $\underline{C}_1$, $u(p) = u_0$ on $\underline{C}_0$ and $u(p) = u_1$ on $\underline{C}_1$, $u_0 \neq u_1$. Then

(1) $$c = \frac{1}{4\pi(u_0 - u_1)} \int \frac{\partial u}{\partial n_i} ds$$

is called the capacity; ds is the element of the line of integration which is in $\underline{C}_1$ and surrounds $\underline{C}_0$; n_i is the interior normal. The quantity (1) represents the capacity per unit length of a cylindrical condenser whose cross section is bounded by the curves $\underline{C}_0$ and $\underline{C}_1$; for this "logarithmic capacity", cf. 32, pp. 333-335.

The analogues of both principles 2.2 (2) and 2.2 (6) hold in this case. (In (2) u_0^{-2} has to be replaced by $(u_0 - u_1)^{-2}$.)

The inner radius r_a and the outer radius $\bar{r}$ are connected with suitable limiting cases of this condenser problem. If $\underline{C}_1 = \underline{C}$ is a given curve, a an interior point and $\underline{C}_0$ a circle of radius ε surrounding a, the capacity $c = c(\varepsilon)$ of this condenser satisfies the equation

$$(2) \qquad \lim_{\varepsilon\to 0} \left(\log \frac{1}{\varepsilon} - \frac{1}{2c(\varepsilon)}\right) = \log \frac{1}{r_a} \quad .$$

On the other hand let $\underline{C}_0 = \underline{C}$ be a given curve and $\underline{C}_1$ a circle of radius ω with fixed center; denoting the capacity of this condenser by $c_1 = c_1(\omega)$ we have

$$(3) \qquad \lim_{\omega\to\infty} \left(\log \omega - \frac{1}{2c_1(\omega)}\right) = \log \bar{r} \ .$$

Equations (2) and (3) follow easily by using the Green functions of the interior and exterior domain (cf. loc. cit.).

2.4. Proof of the principle of Dirichlet. The formal proof of Dirichlet's principle 2.2 (2) is well known. First we observe that, inserting the harmonic function f(p) = u(p) in 2.2 (2) we obtain for the right-hand expression, by Green's formula,

$$(1) \qquad (4\pi)^{-1} u_0^{-2} \iiint |\text{grad } u|^2 \, d\tau = -(4\pi)^{-1} u_0^{-2} \iint u \frac{\partial u}{\partial n} \, d\sigma$$

the last integral being extended over $\underline{A}$; the normal is exterior to $\underline{A}$. (The same integral extended over a large sphere tends to zero.) This expression is

$$= -(4\pi)^{-1} u_0^{-2} \cdot u_0 \iint \frac{\partial u}{\partial n} \, d\sigma = C ,$$

in accordance with 2.1 (2).

Now we write for the proof $f = u + h$, $h = 0$ on $\underline{A}$ and at infinity. Hence

$$(2) \qquad \iiint |\text{grad } f|^2 \, d\tau = \iiint |\text{grad } u|^2 \, d\tau + 2 \iiint \text{grad } u \cdot \text{grad } h \, d\tau + \iiint |\text{grad } h|^2 \, d\tau \ .$$

The first integral on the right-hand side is $4\pi u_0^2 C$, the third is non-negative. The second is by Green's formula

$$(3) \qquad \iiint \text{grad } u \cdot \text{grad } h \, d\tau = - \iint h \frac{\partial u}{\partial n} \, d\sigma - \iiint h \nabla^2 u \, d\tau$$

and both integrals vanish. This proves the assertion.

2.5. Dirichlet's principle with prescribed level surfaces. In order to make a definite use of the inequality 2.2 (2) we must choose a specified function f(p). Although we do not know yet which function f(p) deserves to be preferred, we may have intuitive reasons of various kinds (see 1.29) to choose a certain family of surfaces as level surfaces of f(p). Having made this choice, we have the problem: of all functions f(p) having the given family as level surfaces, select the one for which the right-hand side of 2.2 (2) attains its minimum.

In order to solve this problem, let

(1) $$\psi(p) = \nu$$

be the equation of the given level surfaces. We denote the surface (1) also by $\underline{A}(\nu)$; we assume here that the parameter ν varies from 0 to ∞. Hence (1) represents for $\nu = 0$ the surface $\underline{A}$ and for large ν a large closed (almost spherical) surface. Any function f(p) having the level surfaces (1) must have the form

(2) $$f(p) = \lambda(\psi(p))$$

where $\lambda(t)$ is an arbitrary monotonic function defined between 0 and ∞, $\lambda(0) = u_0$, $\lambda(\infty) = 0$. Also

(3) $$\text{grad } f = \lambda'(\psi) \text{ grad } \psi$$

and so 2.2 (2) takes the form

(4) $$C \leqq (4\pi)^{-1} (\lambda(0))^{-2} \iiint [\lambda'(\psi(p))]^2 |\text{grad } \psi|^2 \, d\tau \quad .$$

The volume element between $\underline{A}(\nu)$ and $\underline{A}(\nu + d\nu)$, $d\nu > 0$, can be written as follows:

$$d\tau = d\sigma \cdot dn = d\sigma \cdot |\text{grad } \psi|^{-1} d\nu$$

where $d\sigma$ is the surface element of $\underline{A}(\nu)$ and $dn = |\text{grad } \psi|^{-1} d\nu$ is the piece of the normal between the two surfaces. This gives for the corresponding part of the integral

$$(\lambda'(\nu))^2 d\nu \iint_{\underline{A}(\nu)} |\text{grad } \psi| \, d\sigma = (\lambda'(\nu))^2 d\nu \iint_{\underline{A}(\nu)} \frac{d\nu}{dn} \, d\sigma \; .$$

The function

(5) $$T(\nu) = \frac{1}{4\pi} \iint_{\underline{A}(\nu)} \frac{d\nu}{dn} \, d\sigma = \frac{1}{4\pi} \iint_{\underline{A}(\nu)} |\text{grad } \psi| \, d\sigma$$

depends only on the set of surfaces $\underline{A}(\nu)$. Consequently, by (4)

(6) $$C \leqq [\lambda(0)]^{-2} \int_0^\infty [\lambda'(\nu)]^2 \, T(\nu) \, d\nu \quad .$$

By Schwarz's inequality

$$(7)\quad \begin{aligned}(\lambda(0))^2 &= \Big(\int_0^\infty \lambda'(\nu)\,d\nu\Big)^2 = \Big\{\int_0^\infty \lambda'(\nu)[T(\nu)]^{1/2}\cdot[T(\nu)]^{-1/2}\,d\nu\Big\}^2 \\ &\leqq \int_0^\infty [\lambda'(\nu)]^2\,T(\nu)\,d\nu \cdot \int_0^\infty [T(\nu)]^{-1}\,d\nu\end{aligned}$$

with equality if $\lambda'(\nu) = \text{const.}\,(T(\nu))^{-1}$. We choose

$$(8)\qquad \lambda(\nu) = \int_\nu^\infty [T(\nu)]^{-1}\,d\nu$$

so that

$$(9)\qquad C \leqq 1 \Big/ \int_0^\infty [T(\nu)]^{-1}\,d\nu \ .$$

This is the best upper estimate that Dirichlet's principle can yield for C if we prescribe the level surfaces.

Similar inequalities hold for condensers and for the two-dimensional case.

2.6. Proof of the principle of Thomson. We give a formal proof for the Thomson principle 2.2 (6). Inserting

$$\vec{f}(p) = -\,\text{grad}\ u$$

in 2.2 (4) and 2.2 (6), where u is defined as in 2.1 ($u = u_0$ on $\underline{A}$), we observe first that, by 2.1 (2),

$$Q = -\frac{1}{4\pi}\iint \frac{\partial u}{\partial n}\,d\sigma = C\,u_0$$

and

$$(4\pi Q^2)^{-1}\iiint |\text{grad}\ u|^2\,d\tau = -(4\pi Q^2)^{-1}\iint u\frac{\partial u}{\partial n}\,d\sigma$$

$$= (4\pi)^{-1}\,(C\,u_0)^{-2}\,u_0\cdot 4\pi u_0\,C = 1/C \ .$$

Now we write for the proof $\vec{f} = -\,\text{grad}\ u + \vec{h}$, so that $\text{div}\ \vec{h} = 0$ and

$$\iint \vec{h}_n\,d\sigma = 0 \ .$$

Then

$$\iiint |\vec{f}|^2 \, d\tau = \iiint |\text{grad } u|^2 \, d\tau - 2 \iiint \text{grad } u \cdot \vec{h} \, d\tau + \iiint |\vec{h}|^2 \, d\tau \quad .$$

The second integral on the right-hand side is by Green's formula

$$= \iiint \left(\frac{\partial u}{\partial x} \vec{h}_x + \frac{\partial u}{\partial y} \vec{h}_y + \frac{\partial u}{\partial z} \vec{h}_z \right) d\tau = \iiint \left(\frac{\partial}{\partial x} (u\vec{h}_x) + \frac{\partial}{\partial y} (u\vec{h}_y) + \frac{\partial}{\partial z} (u\vec{h}_z) \right) d\tau$$

$$= - \iint \left(u\vec{h}_x \cos(n,x) + u\vec{h}_y \cos(n,y) + u\vec{h}_z \cos(n,z) \right) d\sigma = - \iint u\vec{h}_n \, d\sigma = 0 \; .$$

(The surface integral is extended over A and over a very large level surface of u.) This furnishes the assertion.

2.7. Thomson's principle with prescribed lines of force. In order to make a definite use of the inequality 2.2 (6) we must choose a specified vector function $\vec{f}(p)$. Although we do not know yet which function $\vec{f}(p)$ deserves to be preferred, we may have intuitive reasons of various kinds to choose a certain set of lines of force. Again these lines do not determine the vector function $\vec{f}(p)$ completely, only its direction. We form in the usual way infinitesimal tubes of lines starting out from the conductor A and going to infinity. Let dS_0 be the orthogonal cross-section of such a tube at a point p on A and dS the orthogonal cross-section at a distance l measured along the line of force, $0 \leqq l < \infty$. Then in obvious notation

(1) $$|\vec{f}|_0 \, dS_0 = |\vec{f}| \, dS \quad .$$

Also $dS_0 = d\sigma \cos \lambda$ where $d\sigma$ is the surface element of A and λ denotes the angle between the normal of A at the point p and the line of force starting out from p; $0 \leqq \lambda < \pi/2$. Hence prescribing the lines of force and the point function $|\vec{f}|_0$ on A the vector function $\vec{f}(p)$ will be completely determined. Also it will be a sourceless vector. Having chosen the lines of force, we have the problem: for which function $|\vec{f}|_0$ will the right-hand side of 2.2 (6) attain its minimum?

The part of the integral in 2.2 (6) corresponding to the element of the tube between l and l + dl, is

$$|\vec{f}|^2 \, d\tau = |\vec{f}|^2 \, dS \cdot dl = |\vec{f}|_0^2 \, \frac{(d\sigma \cdot \cos \lambda)^2}{dS} \, dl \quad .$$

The ratio $d\sigma/dS$ is a function of l so that the integral

(2) $$L(p) = 4\pi \int_0^\infty \frac{d\sigma}{dS} \, dl$$

is a definite point function on A completely determined by the given geometrical configuration of the lines of force. We have then for the integral in 2.2 (6)

(3) $$(4\pi)^{-1} \iint_{\underline{A}} |\vec{f}|_0^2 \cos^2 \lambda \cdot L(p) \, d\sigma .$$

Also, see 2.2 (4),

(4) $$Q = (4\pi)^{-1} \iint_{\underline{A}} |\vec{f}|_0 \cos \lambda \, d\sigma$$

so that, substituting (3) and (4) in 2.2 (6), we obtain

(5) $$1/C \leqq \iint_{\underline{A}} |\vec{f}|_0^2 \cos^2 \lambda \cdot L(p) \, d\sigma \Big/ \left(\iint_{\underline{A}} |\vec{f}|_0 \cos \lambda \, d\sigma \right)^2 .$$

We choose $|\vec{f}|_0 \cos \lambda = [L(p)]^{-1}$ so that

(6) $$C \geqq \iint_{\underline{A}} [L(p)]^{-1} \, d\sigma .$$

This is the best lower estimate that the principle of Thomson can yield for C, if we prescribe the lines of force as defined above.

Similar inequalities hold for a condenser and for the two-dimensional case.

[2.7 A. Conductance. The two opposite estimates 2.5 (9) and 2.7 (6) for the capacity have interesting connections both in physics and in the theory of conformal mapping. In order to explain these, we need the concept of conductance.

(a) Definition. We regard the surface $\underline{A}$ of the solid body $\underline{B}$ as consisting of three portions $\underline{A}_0$, $\underline{A}_1$ and $\underline{A}'$ without common points. The portion $\underline{A}'$ may be lacking, but $\underline{A}_0$ and $\underline{A}_1$ are closed not-empty sets with appropriately smooth boundary lines. We call $\underline{A}_0$ the "entrance", $\underline{A}_1$ the "exit", and $\underline{A}'$ the "adiabatic" or "impervious" portion of the boundary of $\underline{B}$.

We assume the existence of a harmonic function u satisfying $\nabla^2 u = 0$ throughout the interior of $\underline{B}$, and such that

(1) $$u = u_0 \quad \text{on } \underline{A}_0 , \qquad u = u_1 \quad \text{on } \underline{A}_1 , \qquad \frac{\partial u}{\partial n} = 0 \quad \text{on } \underline{A}' ,$$

$u_0 \neq u_1$. We call

(2) $$C = \frac{1}{4\pi(u_0 - u_1)} \iint \frac{\partial u}{\partial n} \, d\sigma$$

the conductance of B; it is understood that the boundary of B is divided into three specified parts. The integral in (2) is extended over A_1 and n is the interior normal; or the integral is extended over A_0 and n is the exterior normal: in both cases the same value results for the conductance C. This value remains unchanged if the roles of A_0 and A as entrance and exit are interchanged.

If B is a "hollow" solid, contained between an exterior closed surface A_1 and an interior closed surface A_0 which jointly form the full boundary of B (the adiabatic portion is lacking), the conductance defined by (2) reduces to the capacity defined by 2.2 (7). In any case, the quantity C defined by (2) is, except for a factor depending on the physical nature of the homogeneous matter filling B, a thermal, or an electric, conductance.

For example, let B denote a right cylinder, d its altitude and A the area of its base. We consider one of the two parallel bases of B as entrance, the other as exit and the lateral surface as adiabatic. Then u is a linear function of the coordinates, and the conductance of B is

$$C = \frac{A}{4\pi d} \quad . \tag{3}$$

The conductance of a plane domain D may be analogously defined. The definition supposes that the boundary of D is divided into three parts (entrance, exit, and adiabatic portion) in some definite way. The definition, of which we need not discuss the details, reduces in the case of a ring-shaped domain to 2.3 (1) if the boundary has no adiabatic portion. In any case, the conductance of a plane domain D is proportional to the thermal, or electric, conductance of a homogeneous plate, or right cylinder with negligible height, of which the two bases, congruent with D, are adiabatic. For example, let D denote a rectangle, d the length of its height, and a the length of its base; take the two sides of length a as entrance and exit, and the other two sides of length d as adiabatic. Then the conductance is

$$c = \frac{a}{4\pi d} \quad . \tag{4}$$

The conductance of a plane domain remains invariant under conformal mapping. Consider any simply connected domain D, divide its boundary curve C by four points into four arcs, and choose two non-contiguous arcs among these four as entrance and exit. The conductance c of this domain D is equal to the conductance (4) of a suitable rectangle onto which D is mapped conformally; $4\pi c$ is usually called the "module" of D in the theory of conformal mapping.

(b) Characterization by a minimum property. Let f be a function defined in the interior and on the boundary of B, sufficiently smooth and satisfying the boundary conditions

(5) $f = u_0$ on $\underline{A}_0$, $f = u_1$ on $\underline{A}_1$,

but arbitrary otherwise; $u_0 \neq u_1$. Notice that f is free of any specific condition on the adiabatic portion $\underline{A}'$ of the boundary of $\underline{B}$. The conductance, defined by (2), satisfies the inequality

(6) $$C \leq (4\pi)^{-1} (u_0 - u_1)^{-2} \iiint |\text{grad } f|^2 \, d\tau$$

where the integration is extended over the interior of $\underline{B}$. This inequality characterizes C as the minimum of the right-hand side of (6) under the conditions (5).

Only minor modifications of the derivation of Dirichlet's principle given in 2.4 are needed to prove (6). First, we use Green's formula (in almost the same manner as in 2.4 (1)) and the boundary conditions (1) (different from those considered in 2.4) to show that equality is attained in (6) if f coincides with u, the harmonic function arising in the definition (2) of C. Then we put (as in 2.4) $f = u + h$. We conclude from (1) and (5) that

(7) $h = 0$ on $\underline{A}_0$ and $\underline{A}_1$.

The desired inequality (6) will follow from 2.4 (2) if we succeed in showing that the middle term on the right-hand side vanishes. We can show this, however, by 2.4 (3) in taking into account (1), (7) and the fact that $\nabla^2 u = 0$ in the interior of $\underline{B}$.

(c) Estimate on physical grounds. The definition (2) of the conductance applies only to a homogeneous conductor in each point of which the specific conductivity is the same. In the present subsection, however, we take for granted the concept of conductance of an inhomogeneous conductor in which the specific conductivity varies from point to point.

We accept as obvious the principle: if the specific conductivity of any portion of the conductor is diminished or increased, whereas the remainder is unchanged, the conductance of the whole conductor will be diminished or increased, correspondingly, the part and the whole changing in the same sense. We shall apply this principle in the two extreme cases, in diminishing the conductivity to 0 or increasing it to ∞, that is, substituting for the portion in question a "perfect insulator" or a "perfect conductor", respectively. We shall consider especially infinitely thin sheets of perfectly insulating or perfectly conducting material.

Let us examine a homogeneous "hollow" conductor $\underline{B}$, contained between an interior closed surface $\underline{A}_0$ and an exterior closed surface $\underline{A}_1$. We make these surfaces to entrance and exit (in the meaning defined in (a)) by applying to them electrodes of perfectly conducting material. The electric current between these electrodes encounters a resistance proportional to $1/C$ where

C is the conductance, which coincides numerically with the electrostatic capacity of a condenser formed with the surfaces $\underline{A}_0$ and $\underline{A}_1$, as we have observed under (a). We wish to find upper and lower bounds for C.

First, we imagine a system of n + 1 closed surfaces, in a definite sequence, beginning with $\underline{A}_0$ and ending with $\underline{A}_1$; each surface surrounds the preceding. These surfaces divide the hollow conductor $\underline{B}$ into n hollow conductors. Each dividing surface is the exit of the conductor that it delimits from outside and the entrance to the other that it delimits from inside. Let $C_1, C_2, \ldots, C_n$ denote the conductances of these conductors. Take the n - 1 dividing surfaces between $\underline{A}_0$ and $\underline{A}_1$ as infinitely thin sheets of perfectly conducting material. By the principle stated, the introduction of these sheets <u>diminishes the resistance</u> which was 1/C originally; it changes $\underline{B}$ into a series of conductors the resistance of which is the sum of the resistances of the conductors taken separately:

$$\frac{1}{C_1} + \frac{1}{C_2} + \cdots + \frac{1}{C_n} \leqq \frac{1}{C} \quad . \tag{8}$$

If each perfectly conducting surface coincides with some level surface of the original electric current, their introduction does not disturb the flow of electricity and equality is attained in (8).

Second, we imagine a system of open (ring-shaped) surfaces, each bounded by two lines, one on $\underline{A}_0$ and the other on $\underline{A}_1$. Some of these surfaces intersect; we take them as infinitely thin sheets of perfectly insulating material. They divide the conductor $\underline{B}$ into tubes or tubular conductors whose conductances we denote now by $C_1, C_2, \ldots, C_n$. The entrance of each such tubular conductor is a portion of $\underline{A}_0$, its exit a portion of $\underline{A}_1$ and the remainder of the surface is adiabatic, consisting of portions of the perfectly insulating dividing sheets. By the principle stated, the introduction of these sheets <u>diminishes the conductance</u> which was C originally; it changes $\underline{B}$ into a compound conductor the conductance of which is the sum of the conductances of the (parallel) component conductors:

$$C_1 + C_2 + \cdots + C_n \leqq C \quad . \tag{9}$$

If the perfectly insulating surfaces consist entirely of lines of flow of the original electric current, their introduction does not disturb the flow of electricity and equality is attained in (9).

The idea of the foregoing derivation is due to Lord Rayleigh; compare <u>26</u>, vol. 2, pp. 175-189. We followed Maxwell's presentation; see <u>15</u>, vol. 1, pp. 390-397. The inequalities analogous to (8) and (9) in two-dimensional potential theory were found by H. Grötzsch.*

---

* H. Grötzsch, Über einige Extremalprobleme der konformen Abbildung, Berichte d. math.-phys. Klasse der sächsischen Akademie der Wissenschaften, vol. 80 (1928) pp. 367-376.

(d) Proof. Both (8) and (9) follow from the characterization of the conductance by a minimum property, expressed by (6).

The proof of (9) is shorter. Let T_j denote one of the tubular solids into which $\underline{B}$ is divided, C_j its conductance and u the function harmonic in $\underline{B}$ that yields equality in (6); it satisfies (1) (there is no $\underline{A}'$ in the present case). In applying (6) to T_j (instead of $\underline{B}$) in order to estimate C_j, we are allowed to take the u just mentioned for f, and we obtain

$$C_j \leqq (4\pi)^{-1} (u_0 - u_1)^{-2} \iiint_{T_j} |\text{grad } u|^2 \, d\tau \ .$$

Substituting 1, 2, 3, ..., n for j and adding, we prove (9).

In order to prove (8) let H_j denote one of the hollow solids into which $\underline{B}$ is divided, and C_j its conductance; H_j is surrounded by H_{j+1} for j = 1, 2, ..., n-1. We apply (6) to $\underline{B}$. We choose the common boundary of H_j and H_{j+1} as a level surface for f on which $f = f_j$, for j = 1, 2, ..., n-1. We set

$$(10) \qquad f_0 = u_0 \ , \qquad f_n = u_1 \ ,$$

but reserve the choice of $f_1, f_2, \ldots, f_{n-1}$. We choose f as harmonic in each single domain H_j, for j = 1, 2, ..., n, but f need not be harmonic in $\underline{B}$. With this choice

$$(11) \qquad \iiint_{H_j} |\text{grad } f|^2 \, d\tau = 4\pi \, (f_j - f_{j-1})^2 \, C_j \ ,$$

see (b). In view of (10) and (11), (6) yields, applied to $\underline{B}$,

$$(12) \qquad C \leqq (f_0 - f_n)^{-2} \sum_{j=1}^{n} (f_j - f_{j-1})^2 \, C_j \ .$$

By Cauchy's inequality

$$(13) \quad (f_n - f_0)^2 = \left[\sum_{j=1}^{n} (f_j - f_{j-1}) \right]^2 \leqq \sum_{j=1}^{n} (f_j - f_{j-1})^2 \, C_j \sum_{j=1}^{n} 1/C_j \ .$$

We choose $f_1, \ldots, f_{n-1}$ so that equality is attained in (13) (that is, $f_j - f_{j-1}$ proportional to $1/C_j$). This choice gives the most advantageous value to the right-hand side of (12) and transforms (12) into the desired (8).

These proofs show also, with a little additional discussion, that the physically obvious cases of equality in (8) and (9), which we pointed out above, are the only cases of equality. The proof applies to the two-dimensional analogues of (8) and (9) found by Grötzsch. Grötzsch's proof, however, which makes essential use of conformal representation, does not apply so immediately to three dimensions.

(e) Limiting cases. Letting n tend to ∞ in (8) and (9), we can derive important estimates for the capacity which we found in the foregoing by different methods. In view of our former proofs, we content ourselves here with intuitive sketches.

We are given a family of closed surfaces $\underline{A}(\nu)$ in the conductor $\underline{B}$ depending on a continuously varying parameter ν where $0 \leq \nu \leq \nu_1$. The surface $\underline{A}(0)$ coincides with $\underline{A}_0$, $\underline{A}(\nu_1)$ with $\underline{A}_1$ and $\underline{A}(\nu)$ is surrounded by $\underline{A}(\nu')$ if $\nu < \nu'$. Jointly, the surfaces of the family fill the conductor $\underline{B}$. We consider the limiting case of (8) in which $\underline{B}$ is divided into infinitely thin shells by the surfaces $\underline{A}(\nu)$. Let $d\sigma$ denote the surface element of $\underline{A}(\nu)$ and dn the normal to it ending on $\underline{A}(\nu + d\nu)$. Applying (3) to each surface element of the infinitely thin shell between $\underline{A}(\nu)$ and $\underline{A}(\nu + d\nu)$ (conceiving it as a condenser with infinitely near coatings) we compute its conductance (capacity):

$$\iint_{\underline{A}(\nu)} \frac{d\sigma}{4\pi\, dn} = \frac{1}{4\pi\, d\nu} \iint_{\underline{A}(\nu)} \frac{d\nu}{dn}\, d\sigma = \frac{T(\nu)}{d\nu} \;; \tag{14}$$

$T(\nu)$ is an abbreviation (its meaning is the same as in 2.5 (5)). The left-hand side of (8) becomes an integral the element of which is the resistance of an infinitely thin shell, the reciprocal of the conductance (14):

$$\int_0^{\nu_1} \frac{d\nu}{T(\nu)} \leq \frac{1}{C} \,. \tag{15}$$

If the outer surface $\underline{A}_1$ of $\underline{B}$ becomes an infinitely large sphere and $\nu_1 = \infty$ we obtain 2.5 (9). Of course, (15) can be also derived from Dirichlet's principle.

We are given a family of lines depending on two continuously varying parameters. Each line starts from a point p of the surface $\underline{A}_0$ and ends at a point of $\underline{A}_1$; two different lines have no common point. Jointly, the lines of the family fill the conductor $\underline{B}$. We consider the limiting case of (9) in which $\underline{B}$ is divided into infinitely narrow tubes the walls of which consist of lines of the given family. Consider the tube that starts at the point p of the inner surface $\underline{A}_0$. Let l_1 denote its total length (from its initial point to its endpoint on $\underline{A}_1$), dl an element of its length, dS its cross-section perpendicular to dl and $d\sigma$ the element cut out by the tube from the surface $\underline{A}_0$. Applying (3) to each longitudinal element of the tube (considering it as a thin wire with variable cross-section) we compute its resistance (the reciprocal of its conductance):

$$\int_0^{l_1} \frac{4\pi\, dl}{dS} = \frac{4\pi}{d\sigma} \int_0^{l_1} \frac{d\sigma}{dS}\, dl = \frac{L(p)}{d\sigma} \quad ; \tag{16}$$

L(p) is an abbreviation (which goes over into 2.7 (2) if $l_1 = \infty$). The left-hand side of (9) becomes an integral extended over the inner surface of B the element of which is the conductance of an infinitely narrow tube:

$$\iint_{A_0} \frac{d\sigma}{L(p)} \leqq C \; . \tag{17}$$

If the outer surface A_1 of B becomes an infinitely large sphere so that $l_1 = \infty$ in (16), we obtain 2.7 (6). Of course, (17) can be derived from Thomson's principle also in the slightly more general case considered in the present section.]

2.8. The principle of Gauss. Under this term we refer to a special case of the Thomson principle which is often of importance.*

We consider a charge distribution on A defined by the density function μ; the total charge is denoted by Q so that

$$\iint_A \mu\, d\sigma = Q \; . \tag{1}$$

Let $Q \neq 0$. We denote the potential of this charge by V and choose in Thomson's principle 2.2 (6)

$$\vec{f}(p) = \text{grad } V \; . \tag{2}$$

Then conditions 2.2 (3) and (4) are satisfied and we have by Green's formula

$$\iiint_E |\text{grad } V|^2\, d\tau = - \iint_A V \frac{\partial V}{\partial n_e}\, d\sigma \quad , \tag{3}$$

n_e being the exterior normal of A. The volume integral is extended over the exterior E of A; the surface integral over a large spherical surface tends to zero. Denoting the interior normal by n_i, we have

* See, for instance, Ph. Frank-R. v. Mises, Die Differential- und Integralgleichungen der Mechanik und Physik, vol. 1, 2. edition, 1930, pp. 772-773.

(4) $$\frac{\partial V}{\partial n_e} + \frac{\partial V}{\partial n_i} = 4\pi\mu$$

so that

$$\iiint_E |\text{grad } V|^2 \, d\tau = - \iint_A V \left(-4\pi\mu - \frac{\partial V}{\partial n_i} \right) d\sigma$$

$$= 4\pi \iint_A V\mu \, d\sigma + \iint_A V \frac{\partial V}{\partial n_i} \, d\sigma$$

$$= 4\pi \iint_A V\mu \, d\sigma - \iiint |\text{grad } V|^2 \, d\tau$$

the last integration being extended over the <u>interior</u> of $\underline{A}$. Consequently

$$\iiint_E |\text{grad } V|^2 \, d\tau \leqq 4\pi \iint_A V\mu \, d\sigma$$

and

(5) $$1/C \leqq Q^{-2} \iint_A V\mu \, d\sigma \quad .$$

This is the principle of Gauss which can be stated as follows:

> Let V be the potential of any charge Q distributed over $\underline{A}$ with surface density μ. Then the least value that the quantity on the right-hand side of (5) can attain is $1/C$. This minimum is attained for the electrostatic equilibrium distribution.

More precisely we obtain by the above argument

(6) $$1/C \leqq Q^{-2} \iint_A V\mu \, d\sigma - (4\pi)^{-1} Q^{-2} \iiint |\text{grad } V|^2 \, d\tau \quad ;$$

the volume integral is extended over the <u>interior</u> of $\underline{A}$.

Another form of the integral on the right-hand side of (5) is the following:

(7) $$\iint_A V\mu \, d\sigma = \iiiint \frac{\mu_1 \mu_2}{r_{12}} \, d\sigma_1 \, d\sigma_2 \quad ;$$

here $d\sigma_1$, $d\sigma_2$ are the surface elements at two points running independently of each other over $\underline{A}$; μ_1, μ_2 are the densities at these points and r_{12}

their distance.

The extension of the principle of Gauss to the case of a condenser (see 2.2 (c) and (d)) is obvious. Such a case is dealt with in Chapter IV.

A modification of the previous results which is often useful, follows, in fact, more directly from Thomson's principle. Let A be a given surface. We denote by V the potential of arbitrary charges of total amount Q, $Q \neq 0$, placed in the interior (not on the boundary) of A. Defining $\vec{f}(p)$ by (2), we have again (3) so that 2.2 (6) yields

$$(8) \qquad \frac{1}{C} \leq - \frac{1}{4\pi Q^2} \iint_A V \frac{\partial V}{\partial n_e} \, d\sigma .$$

This inequality will be used in the next section.

2.9. A method of approximation. (a) Let A be a Jordan surface, p_0 an arbitrary point in the interior of A. Denoting by r the distance of a variable point p from the fixed point p_0, we have in $1/r$ a basically important special harmonic function of the "charge" 1. Maxwell's celebrated representation of all spherical harmonics of given degree n can be formulated then as follows:

$$(1) \qquad K \frac{\partial}{\partial l_1} \frac{\partial}{\partial l_2} \cdots \frac{\partial}{\partial l_n} \left(\frac{1}{r}\right)$$

where K is a constant and $l_1, l_2, \ldots, l_n$ denote n arbitrary unit vectors.* The result of this differential operation is an expression of the form $H_n(x,y,z)\, r^{-2n-1}$ where $H_n(x,y,z)$ is the most general homogeneous polynomial of degree n satisfying Laplace's differential equation $\nabla^2 H_n = 0$. For $n \geq 1$, the charge of the harmonic function (1) is $= 0$. Consequently any harmonic function of the form

$$(2) \qquad V_0 = r^{-1} + H_1(x,y,z)r^{-3} + H_2(x,y,z)r^{-5} + \ldots + H_n(x,y,z)r^{-2n-1}$$

can be interpreted as the potential of certain charges concentrated at p_0 with total charge 1. Hence we have by 2.8 (8):

$$(3) \qquad \frac{1}{C} \leq - \frac{1}{4\pi} \iint_A V_0 \frac{\partial V_0}{\partial n_e} \, d\sigma .$$

* Cf., for instance, R. Courant - D. Hilbert, Methoden der mathematischen Physik, 1924, vol. 1, p. 423.

We use now the following analogue of a fundamental theorem of Runge:*

> Let V be a harmonic function regular outside of a surface $\underline{A}'$ which is contained entirely in the interior of $\underline{A}$. Let the charge of V be equal to 1. If ε is a given positive number we can find a special harmonic function V_0 of the form (2) such that on $\underline{A}$:

$$(4) \qquad |V - V_0| < \varepsilon , \qquad |\text{grad } V - \text{grad } V_0| < \varepsilon .$$

In the special case when $\underline{A}$ is an <u>analytic surface</u>, the potential V of the electrostatic equilibrium distribution satisfies the condition of this theorem. Indeed in this case V can be continued analytically into the interior of $\underline{A}$. From this we conclude, in view of (3), that 1/C is the <u>greatest lower bound</u> of the quantities

$$-\frac{1}{4\pi} \iint_{\underline{A}} V_0 \frac{\partial V_0}{\partial n_e} \, d\sigma = \frac{1}{4\pi} \iiint_{E} |\text{grad } V_0|^2 \, d\tau$$

(the integration extended over the exterior E of $\underline{A}$) where V_0 is an arbitrary function of the special type (2).

The same conclusion can be drawn if $\underline{A}$ is not necessarily analytic but star-shaped with respect to an interior point p_0. Indeed we have then, if $0 < \lambda < 1$,

$$\frac{1}{(1-\lambda)C} = -\frac{1}{4\pi} \iint_{\underline{A}_\lambda} V' \frac{\partial V'}{\partial n_e} \, d\sigma$$

where $\underline{A}_\lambda$ is the surface arising from $\underline{A}$ by contraction with respect to p_0 in the ratio $1 - \lambda : 1$; V' denotes the potential of the electrostatic equilibrium on $\underline{A}_\lambda$, the charge of V' being 1. Hence, if E_λ and E denote the domains exterior to $\underline{A}_\lambda$ and $\underline{A}$, respectively, we have

$$\frac{1}{(1-\lambda)C} = \frac{1}{4\pi} \iiint_{E_\lambda} |\text{grad } V'|^2 \, d\tau$$

$$> \frac{1}{4\pi} \iiint_{E} |\text{grad } V'|^2 \, d\tau = -\frac{1}{4\pi} \iint_{A} V' \frac{\partial V'}{\partial n_e} \, d\sigma$$

---

* See Frank-Mises, loc. cit., pp. 760-762. Cf. also J. L. Walsh, The approximation of harmonic functions by harmonic polynomials and by harmonic rational functions, Bulletin of the American Mathematical Society, vol. 35 (1929) pp. 499-544; in particular, pp. 535-541. - The inequalities (4) arise from the results quoted by an inversion.

Now we apply the approximation theorem to V' on $\underline{A}$. The last integral differs by an arbitrarily small amount from an integral in which V' is replaced by an appropriate V_0. This establishes the assertion.

(b) The last remark leads to the following algorithm for the computation of the capacity C. The most general harmonic polynomial $H_n(x,y,z)$ contains 2n + 1 arbitrary constants and the expression

$$\frac{1}{4\pi}\iiint\left|\operatorname{grad}\sum_{\nu=0}^{n}\frac{H_\nu(x,y,z)}{r^{2\nu+1}}\right|^2 d\tau \tag{5}$$

is a quadratic form of these constants whose total number is

$$N = \sum_{\nu=0}^{n}(2\nu+1) = (n+1)^2 \quad .$$

Let us denote the minimum of this form under the condition $H_0 = 1$ by m_n. Then

$$m_0 \geqq m_1 \geqq m_2 \geqq \cdot\cdot\cdot \geqq m_n \geqq \cdot\cdot\cdot \tag{6}$$

and

$$\lim_{n\to\infty} m_n = 1/C \quad . \tag{7}$$

The quantities m_n can be computed in the following well-known manner. We denote by $k_\mu(x,y,z)$, $\mu = 1,2,\ldots,$ N, any system of independent spherical harmonics of degree -h, h = 1,2,..., n + 1; here $k_1 = 1/r$ and we assume that k_μ is without charge for $\mu > 1$. We denote by c_μ arbitrary real constants. Then the minimum m_n of the quadratic form

$$\frac{1}{4\pi}\iiint\left|\operatorname{grad}\sum_{\mu=1}^{N} c_\mu k_\mu(x,y,z)\right|^2 d\tau = \sum_{i=1}^{N}\sum_{j=1}^{N}\alpha_{ij}\, c_i\, c_j ,$$

$$\begin{aligned}\alpha_{ij} = \alpha_{ji} &= \frac{1}{4\pi}\iiint \operatorname{grad} k_i \cdot \operatorname{grad} k_j \, d\tau \\ &= -\frac{1}{4\pi}\iint_{\underline{A}} k_i \frac{\partial k_j}{\partial n_e}\, d\sigma = -\frac{1}{4\pi}\iint_{\underline{A}} k_j \frac{\partial k_i}{\partial n_e}\, d\sigma,\end{aligned} \tag{8}$$

under the condition $c_1 = 1$, is the quotient of two determinants:

(9) $$m_n = [\alpha_{ij}]_1^N \Big/ [\alpha_{ij}]_2^N \quad .$$

We have, for instance,

(10) $$m_0 = -\frac{1}{4\pi} \iint\limits_{\underline{A}} \frac{1}{r} \frac{\partial(1/r)}{\partial n_e} \, d\sigma \quad .$$

In the special case when the given surface A admits of a certain finite symmetry group, that is, a group of rotations and rotatory reflections about p_0, the number N of the variables in this algorithm can be reduced. Indeed, it is obvious that the minimum of (5) is attained for functions V_0 which admit this group of symmetry, that is, which are invariant under the transformations of this group. Thus it is sufficient to consider only the linear combination V_0 of the spherical harmonics admitting the given group of symmetry.

We repeat that this algorithm has been established under the conditon that A is either analytic or star-shaped.

2.10. The two-dimensional case. The analogue of the principle of Gauss in the plane is the following inequality for the outer radius $\bar{r}$ of a curve C:

(1) $$\log \frac{1}{\bar{r}} \leq \frac{1}{q^2} \int\limits_{\underline{C}} V \mu \, ds \quad .$$

Here μ is the linear density of an arbitrary distribution over C with the total charge q, $q \neq 0$ and V is the logarithmic potential of this charge,

(2) $$\int\limits_{\underline{C}} \mu \, ds = q \ , \qquad V = \int\limits_{\underline{C}} \log \frac{1}{r} \cdot \mu \, ds \ ,$$

ds is the arc-element of C. The left hand side in (1) is the minimum of the quantity on the right hand side for all possible distributions over C.

2.11. A remark on discrete masses. Returning again to Gauss' principle in space, let us consider n distinct points $p_1, p_2, \ldots, p_n$ on the surface A. We concentrate the charge $1/n$ at each of these points, so that the total charge is 1. The analogue of the inequality 2.8 (5) (cf. also 2.8 (7)) would be in this case

(1) $$1/C \leq n^{-2} \sum_{\substack{i,j=1,2,\ldots,n \\ i \neq j}} |p_i p_j|^{-1} \quad .$$

This is, however, false in general. On the contrary, if we denote the minimum of the right hand expression (as p_1, p_2, ..., p_n run independently of each other on the given surface A) by Σ_n we have (see 23, p. 12, (6))

$$(2) \qquad 1/C \geqq \frac{n}{n-1} \quad \Sigma_n > \Sigma_n \quad .$$

We have $\lim_{n \to \infty} \Sigma_n = 1/C$.

A similar remark holds in the two-dimensional case, the treatment of which by M. Fekete (6) suggested the study of the quantities Σ_n.

2.12. Another way of estimating capacities. (a) Let A be an arbitrary surface. We consider an arbitrary distribution of charges in the interior of A with total charge 1 and with the potential V outside of A. Then the inequalities

$$(1) \qquad \min V \leqq 1/C \leqq \max V$$

hold; here the maximum and minimum of V has to be taken on the surface A.

These inequalities are easy to prove. Using the function u which we have introduced in 2.1 (potential of the charge which is in electrostatic equilibrium on A) we assume that $u_0 = 1$ so that $X_0 = C$. Consider now the difference V - u/C. It is "without charge" at infinity, that is, the leading term of its expansion in spherical harmonics has the form Y_n/r^{n+1} where Y_n is a surface harmonic of degree n, $n \geqq 1$. (The only exception is V - u/C $\equiv 0$.) This leading term takes both positive and negative values on any sphere with center at the origin and, therefore, the function V - u/C must do the same on sufficiently large spheres of this kind. Thus, the maximum of this function is > 0, its minimum < 0, and, since its value at infinity is $= 0$, both the maximum and the minimum must be attained on the boundary A, and just this is asserted by (1).

The equality signs in (1) hold only if V = u/C.

These rather crude inequalities (1) furnish in certain cases bounds for C without necessitating the computation (or estimation) of the integrals occuring in the principles of Dirichlet and Thomson. One of the bounds can easily be compared with that given by Gauss' principle: The lower bound for C in 2.8 (5) is obviously higher than that in (1).

(b) A slightly more general theorem is the following. Let C be the capacity of a condenser bounded by two surfaces A_0 and A_1, where A_0 is in the interior of A_1. We denote by v an arbitrary harmonic function defined in the field between these two surfaces. Let

$$(2) \qquad \begin{cases} m_0 \leqq v \leqq M_0 & \text{on } A_0 , \\ m_1 \leqq v \leqq M_1 & \text{on } A_1 . \end{cases}$$

We introduce the flux of v:

$$\Gamma = \frac{1}{4\pi} \iint \frac{\partial v}{\partial n} \, d\sigma \tag{3}$$

where the integration is extended over an arbitrary surface surrounding $\underline{A}_0$ and contained in $\underline{A}_1$; the normal is the exterior one. Then we have the inequalities:

$$m_1 - M_0 \leqq \frac{\Gamma}{C} \leqq M_1 - m_0 \quad . \tag{4}$$

Let u be the harmonic function defined by the conditions: $u = 0$ on $\underline{A}_0$, $u = 1$ on $\underline{A}_1$, and let

$$w = v - \frac{\Gamma}{C} u \quad . \tag{5}$$

Then the flux of w vanishes across any surface $\underline{A}$ containing $\underline{A}_0$ and contained in $\underline{A}_1$ and

$$m_0 \leqq w \leqq M_0 \qquad \text{on } \underline{A}_0 \;,$$

$$m_1 - \frac{\Gamma}{C} \leqq w \leqq M_1 - \frac{\Gamma}{C} \qquad \text{on } \underline{A}_1 \;.$$

We can now restate the conclusion (4) that we have to prove by saying that the intervals (m_0, M_0) and $(m_1 - \Gamma/C, M_1 - \Gamma/C)$ have at least a point in common.

In fact, if this is not so, there is a value μ such that all values taken by w on one of the surfaces, say $\underline{A}_1$, are greater than μ, and all values of w on the other surface, $\underline{A}_0$, less than μ. Therefore, $w = \mu$ on a (level) surface $\underline{A}$ surrounding $\underline{A}_0$ and surrounded by $\underline{A}_1$. In the space between $\underline{A}_0$ and $\underline{A}$ the maximum of w is μ, attained at all points of $\underline{A}$, and, therefore,

$$\frac{\partial w}{\partial n_e} \geqq 0$$

at all points of $\underline{A}$ and $\partial w / \partial n_e > 0$ in some points (unless w is a constant). If this is so, however, the flux of w across $\underline{A}$ cannot vanish. In order to avoid a contradiction we must admit our assertion which is equivalent to (4).

The analogous case of a condenser defined by separated surfaces can be treated similarly. We shall deal with such a case in 4.5.

CHAPTER III. APPLICATIONS OF THE PRINCIPLES OF DIRICHLET AND THOMSON TO ESTIMATION OF THE CAPACITY

3.1. Introduction. In the following we shall use the two principles discussed in the foregoing chapter in order to obtain upper and lower bounds for the capacity. We apply these principles mainly in the forms 2.5 (9) and 2.7 (6), choosing the level surfaces and lines of force in an appropriate way. We consider three classes of surfaces and make this choice in each case in a different way:

(1) convex surfaces; we choose the exterior parallel surfaces and the normals of the given surface;

(2) surfaces which are star-shaped with respect to a certain interior point p_0; we choose the surfaces similar (with respect to p_0) to the given surface and the rays issuing from p_0;

(3) surfaces of revolution; we choose the surfaces of revolution obtained by rotating the level curves of the exterior conformal mapping of the meridian curve of the given surface onto a circle; and we choose the lines corresponding in this conformal mapping to the radii of the circle.

3.2. The inequality $C \geq \bar{V}$. This inequality of Poincaré-Faber-Szegö was mentioned in 1.12; it can be proved by combining Dirichlet's principle 2.2 (2) and the process of symmetrization with respect to a point.

We assume that f is the minimizing function of Dirichlet's integral satisfying the boundary condition $f = u_0 = 1$ on the given surface $\underline{A}$ and vanishing at infinity. We call $\underline{A}(\nu)$ the surface (level surface) consisting of all the points at which $f = 1 - \nu$.

We symmetrize with respect to a point. We mean by this term the following procedure. The surface $\underline{A}(\nu)$ is replaced by a sphere $\underline{A}^*(\nu)$ of equal volume and with center at a fixed point which we choose as the origin. If $V(\nu)$ stands for the volume enclosed by $\underline{A}(\nu)$ and R is the radius of the sphere $\underline{A}^*(\nu)$, we have $V(\nu) = 4\pi R^3/3$. (This defines $R = R(\nu)$ as a function of ν.) In particular, $\underline{A}^*(0) = \underline{A}^*$ is the sphere whose volume is the same as that of the given surface $\underline{A}$. The domain E exterior to $\underline{A}$ is transformed by this process into the domain E^* exterior to the sphere $\underline{A}^*$.

Moreover, by symmetrizing f, we introduce a new function f^*. As $f = 1 - \nu$

on the surface $\underline{A}(\nu)$, we define f^* by the condition that $f^* = 1 - \nu$ on the sphere $\underline{A}^*(\nu)$. This function f^* has spherical symmetry, that is, it is constant on concentric spheres with center at the origin. It satisfies the boundary conditions $f^* = 1$ on $\underline{A}^*$ and $f^* = 0$ at infinity. Our aim is to show that symmetrization with respect to a point <u>diminishes</u> Dirichlet's integral, or

$$(1) \qquad \iiint_E |\text{grad } f|^2 \, d\tau \geq \iiint_{E^*} |\text{grad } f^*|^2 \, d\tau \quad .$$

Since the left-hand side is $4\pi C$ (C is the capacity of $\underline{A}$) and the right-hand side is greater than $4\pi C^*$ (C^* is the capacity of $\underline{A}^*$) the assertion will follow.

In fact, we shall prove a little more, namely that even any "element" of Dirichlet's integral diminishes. By an element we understand the contribution of the shell-shaped domain bounded by the surfaces $\underline{A}(\nu)$ and $\underline{A}(\nu + d\nu)$ which we wish to compare with the contribution of the spherical shell between the spheres $\underline{A}^*(\nu)$ and $\underline{A}^*(\nu + d\nu)$; we take $d\nu > 0$. The volume of this domain is

$$(2) \qquad V(\nu + d\nu) - V(\nu) = dV = V'(\nu)\, d\nu \quad .$$

Another way of representing the same volume is the following:

$$(3) \qquad dV = \iint d\sigma \cdot dn$$

where the integration is extended over $\underline{A}(\nu)$, $d\sigma$ is the surface element of $\underline{A}(\nu)$ and dn is the piece of the normal of this surface between the surfaces corresponding to ν and $\nu + d\nu$. But $|\text{grad } f| = d\nu/dn$ so that (3) can be written as follows:

$$(4) \qquad V'(\nu) = \iint |\text{grad } f|^{-1} \, d\sigma \quad .$$

We observe also that the contribution of the infinitesimal volume mentioned above to Dirichlet's integral is

$$(5) \qquad \iint |\text{grad } f|^2 \, d\sigma \cdot dn = \iint |\text{grad } f|^2 \, d\sigma \cdot \frac{d\nu}{|\text{grad } f|} = \iint |\text{grad } f| \, d\sigma \cdot d\nu \quad .$$

All integrals are taken on $\underline{A}(\nu)$.

We apply now Schwarz's inequality

$$(6) \qquad \iint |\text{grad } f| \, d\sigma \iint |\text{grad } f|^{-1} \, d\sigma \geq \left(\iint d\sigma\right)^2 = [S(\nu)]^2$$

where $S(\nu)$ is the surface area of $\underline{A}(\nu)$. It follows from (4) and (6) that

$$\iint |\text{grad } f|\, d\sigma \geq [S(\nu)]^2\, [V'(\nu)]^{-1} \tag{7}$$

Using the isoperimetric inequality (1.13 (1))

$$[S(\nu)]^2 \geq (36\pi)^{2/3}\, [V(\nu)]^{4/3} \tag{8}$$

we obtain from (7) that

$$\iint_{\underline{A}(\nu)} |\text{grad } f|\, d\sigma \geq (36\pi)^{2/3}\, [V(\nu)]^{4/3}\, [V'(\nu)]^{-1} \tag{9}$$

The quantity corresponding to (9) for the symmetrized function f^* (the contribution of an infinitesimal spherical shell to Dirichlet's integral formed with f^*) is

$$\begin{aligned} \iint_{\underline{A}^*(\nu)} |\text{grad } f^*|\, d\sigma &= \frac{d\nu}{dR} \cdot 4\pi R^2 = \frac{d\nu}{4\pi R^2 dR} \cdot (4\pi R^2)^2 \\ &= (36\pi)^{2/3}\, [V(\nu)]^{4/3}\, [V'(\nu)]^{-1} \quad . \end{aligned} \tag{10}$$

Here we used the relation $V(\nu) = 4\pi R^3/3$, $V'(\nu)\, d\nu = 4\pi R^2\, dR$.

Comparison of (9) and (10) establishes the assertion (1); see (5).

A similar argument is possible in the case of a condenser and also in the two-dimensional case. The basic ideas of this proof will return in several connections. See for instance 5.9 A and the notes C, D, E, F.

3.3. <u>Mean values of the radii of a solid</u>. Before we proceed with the applications indicated in 3.1 we mention another inequality which follows from 3.2. Let $\underline{A}$ be a surface which is star-shaped with respect to an interior point p_0. We denote by r the radius vector leading from p_0 to a boundary point p of $\underline{A}$ and by $d\omega$ the <u>solid angle</u> subtended at p_0 by the surface element $d\sigma$ at p. Then

$$d\sigma \cos(r, n_e) = r^2\, d\omega \quad . \tag{1}$$

Let λ be any real number. The mean-value

$$N_\lambda = \left(\frac{1}{4\pi} \iint r^\lambda\, d\omega\right)^{1/\lambda} \tag{2}$$

where the integration is extended over the unit sphere, is an <u>increasing</u>

function of λ (as is well known). The following two cases are of particular interest.

(1) $\lambda = -1$. Then

$$(N_{-1})^{-1} = \frac{1}{4\pi}\iint r^{-1}\, d\omega = -\frac{1}{4\pi}\iint_{\underline{A}} \frac{1}{r}\frac{\partial(1/r)}{\partial n_e}\, d\sigma \tag{3}$$

$$= \frac{1}{4\pi}\iiint_{E} \left|\operatorname{grad}\frac{1}{r}\right|^2 d\tau \;;$$

the last integral is extended over the space E outside of $\underline{A}$. This is the quantity m_0 defined in 2.9 (10).

(2) $\lambda = 3$. Then $N_3 = \bar{V}$ since $\frac{1}{3} r \cdot r^2\, d\omega$ is a volume element of the solid bounded by $\underline{A}$.

By the theorem of 3.2

$$C \geqq N_\lambda \qquad \text{for} \qquad \lambda \leqq 3 \;. \tag{4}$$

Thus we observe that the lower bound for the capacity obtained as a first step of the algorithm described in 2.9 (which uses m_0) is never above the lower bound $\bar{V}$.

In 6.11 we shall prove that the inequality (4) ceases to be generally true if $\lambda > 3$.

3.4. Parallel surfaces. (a) Let $\underline{A}$ be a convex surface, $\underline{A}_\nu$ the outer parallel surface of $\underline{A}$ at the distance ν. Then, using the notation of 2.5, we have

$$\frac{d\nu}{dn} = 1 \;, \tag{1}$$

and the quantity $4\pi T(\nu)$, see 2.5 (5), will be simply the surface area $S(\nu)$ of $\underline{A}_\nu$. Consequently, by 2.5 (9),

$$C \leqq \left(4\pi \int_0^\infty \frac{d\nu}{S(\nu)}\right)^{-1} \;. \tag{2}$$

This inequality is due to Szegö (31).

Substituting the well known expression

$$S(\nu) = S + 2M\nu + 4\pi\nu^2 \tag{3}$$

where $S = S(0)$ and M denote the surface area and the Minkowski constant of the given surface $\underline{A}_0$, respectively, we obtain the inequality (1.13 (5))

(4) $$\frac{1}{C} \geq \frac{4\pi}{M} \frac{1}{2\varepsilon} \log \frac{1+\varepsilon}{1-\varepsilon}, \qquad \varepsilon^2 = 1 - \frac{4\pi S}{M^2} = 1 - \left(\frac{\bar{S}}{\bar{M}}\right)^2 .$$

The quantity ε is between 0 and 1 according to the inequality $\bar{S} \leq \bar{M}$ of Minkowski (1.13 (3)); ε might be called the "isoperimetric deficit" and is a measure for the deviation of the given surface $\underline{A}$ from the sphere.

Since the quantity multiplying $4\pi/M$ on the right-hand side of (4) is greater than 1 we find the following inequality (1.13 (3))

(5) $$C \leq \bar{M} .$$

The interpretation of the right-hand expression in (4) in terms of the capacity of a prolate spheroid is immediate. (Cf. 1.13.)

Inequality (2) can be extended to the case of condensers bounded by <u>parallel surfaces</u>. We have to use Dirichlet's principle in the form 2.7 A (15). For the resulting inequality see Pólya-Szegö <u>24</u>, p. 17, (5.22).

In a similar fashion we can prove a corresponding inequality in the plane; this is the case of cylindrical condensers with cross-sections which are parallel curves. For the resulting upper bound, see <u>24</u> (5.23).

(b) We follow now the notation of 2.7. Choosing for the lines of force the exterior normals of $\underline{A}$ we have

(6) $$\lambda = 0, \quad dS_0 = d\sigma, \quad l = \nu .$$

We map the surface $\underline{A}$ by parallel normals onto the unit sphere. Denoting the principal radii of curvature of $\underline{A}$ by R_1 and R_2 and the surface element of the unit sphere by $d\omega$, we have*

(7) $$\begin{cases} d\sigma = R_1 R_2 \, d\omega , \\ dS = (R_1 + \nu)(R_2 + \nu)\, d\omega . \end{cases}$$

Hence, see 2.7 (2),

(8) $$L(p) = 4\pi \int_0^\infty \frac{R_1 R_2 \, d\nu}{(R_1+\nu)(R_2+\nu)} = \frac{4\pi \log (R_1/R_2)}{R_2^{-1} - R_1^{-1}}$$

so that, by 2.7 (6), the bound

(9) $$C \geq \frac{1}{4\pi} \iint_{\underline{A}} \frac{R_2^{-1} - R_1^{-1}}{\log R_2^{-1} - \log R_1^{-1}} \, d\sigma$$

* W. Blaschke, Vorlesungen über Differentialgeometrie, vol. 1, 1921, p. 68 (80); p. 82 (142).

follows.

This inequality can be extended to the case of condensers bounded by the parallel surfaces $\underline{A} = \underline{A}_0$ and $\underline{A}_d$. We have

$$(10) \qquad C \geqq \frac{1}{4\pi} \iint_{\underline{A}} \frac{R_2^{-1} - R_1^{-1}}{\log (1 + R_2^{-1} d) - \log (1 + R_1^{-1} d)} d\sigma .$$

A similar inequality holds for cylindrical condensers; cf. 24, p. 18, (5.23).

3.5. Similar surfaces. (a) Let $\underline{A}$ be a closed surface which is star-shaped with respect to one of its interior points p_0. Enlarging $\underline{A}$ with respect to p_0 in the ratio $1:1 + \nu$, $\nu > 0$, we obtain a surface similar to $\underline{A}$ and similarly situated which we denote by $\underline{A}_\nu$; $\underline{A}_0 = \underline{A}$. As ν varies we obtain a set of similar surfaces which we choose as level surfaces.

Let us compute the corresponding quantity $T(\nu)$; see 2.5 (5). In the transition from $\underline{A}$ to $\underline{A}_\nu$ the ratio $d\nu/dn$ does not change but the surface element is enlarged in the proportion $1 : (1 + \nu)^2$. Hence $T(\nu) = (1 + \nu)^2 T(0)$.

In order to compute $T(0)$, we write the equation of $\underline{A}$ in the form $f(x,y,z) = 0$ and assume that p_0 is the origin. Then the equation of $\underline{A}_\nu$ can be written as follows:

$$(1) \qquad f\left(\frac{x}{1+\nu}, \frac{y}{1+\nu}, \frac{z}{1+\nu}\right) = 0 .$$

Differentiating this with respect to x, y, z, we find for $\nu = 0$

$$(2) \qquad f_x - \nu_x (x f_x + y f_y + z f_z) = 0 , \ \ldots ,$$

$$(3) \qquad \frac{d\nu}{dn} = |\mathrm{grad}\ \nu| = \frac{|\mathrm{grad}\ f|}{x f_x + y f_y + z f_z}$$

provided the gradient is directed outward. The last expression can be written in the form $1/h$ where h is the distance of the origin from the tangent plane at the variable point p of $\underline{A}$. Thus we have

$$(4) \qquad T(0) = \frac{1}{4\pi} \iint_{\underline{A}} \frac{d\sigma}{h}$$

In view of the form of $T(\nu)$ and 2.5 (9) we obtain the inequality

$$(5) \qquad C \leqq \frac{1}{4\pi} \iint_{\underline{A}} \frac{d\sigma}{h}$$

(b) We note that this result is a consequence of the inequality 3.4 (5) if $\underline{A}$ is convex. Indeed by Schwarz's inequality

$$(6)\qquad \left(\iint_{\underline{A}} d\sigma\right)^2 \leqq \iint_{\underline{A}} h\,d\sigma \iint_{\underline{A}} h^{-1}\,d\sigma \ .$$

The first integral on the right-hand side is 3V so that

$$(7)\qquad S^2 \leqq 3V \iint_{\underline{A}} h^{-1}\,d\sigma \ ,$$

or

$$(8)\qquad \frac{\bar{S}^4}{\bar{V}^3} \leqq \frac{1}{4\pi} \iint_{\underline{A}} \frac{d\sigma}{h}$$

But by the quoted result and by Minkowski's second inequality (1.13 (4))

$$(9)\qquad C \leqq \bar{M} \leqq \frac{\bar{S}^4}{\bar{V}^3} \leqq \frac{1}{4\pi} \iint_{\underline{A}} \frac{d\sigma}{h} \ .$$

(c) Let C be the capacity of a condenser bounded by two similar and similarly situated surfaces which we denote by $\underline{A} = \underline{A}_0$ and $\underline{A}' = \underline{A}_{\nu_1}$, and their surface areas by S and S', respectively. Then we obtain as above, from 2.7 A (15),

$$(10)\qquad C \leqq T(0)\left(\int_0^{\nu_1} \frac{d\nu}{(1+\nu)^2}\right)^{-1} = \frac{1+\nu_1}{\nu_1}\,T(0)$$

so that, by (4),

$$(11)\qquad C \leqq \frac{1+\nu_1}{4\pi\nu_1} \iint_{\underline{A}} \frac{d\sigma}{h} \ .$$

We define d as the distance from a variable point of $\underline{A}$ to the tangent plane at the corresponding point of $\underline{A}'$ or, briefly, the distance between corresponding tangent planes. Obviously

$$(d+h)/h = S'^{1/2}/\,S^{1/2} = 1+\nu_1$$

and so we can rewrite inequality (11) in the forms

$$(12)\qquad C \leqq \left\{1 - \left(\frac{S}{S'}\right)^{1/2}\right\}^{-1} \frac{1}{4\pi} \iint \frac{d\sigma}{h} = \left(\frac{S'}{S}\right)^{1/2} \iint \frac{d\sigma}{4\pi d} \ .$$

The last integral could be chosen as an approximation to C on intuitive grounds.

(d) For lower bounds we use 2.7 (6). Choosing as the lines of force the radii issuing from the point p_0, we have

$$(13)\quad \begin{cases} dS = (1+\nu)^2\, d\, S_0 = (1+\nu)^2 \cos(r,n)\, d\,\sigma, & l = \nu r, \\ L(p) = \dfrac{4\pi r}{\cos(r,n)} \displaystyle\int_0^\infty \dfrac{d\nu}{(1+\nu)^2} = \dfrac{4\pi r}{\cos(r,n)} & \end{cases}$$

so that (3.3 (1))

$$(14)\qquad C \geq (4\pi)^{-1} \iint_{\underline{A}} r^{-1} \cos(r,n)\, d\sigma = (4\pi)^{-1} \iint_{\underline{A}} r\, d\omega .$$

This lower bound of C is identical with the quantity denoted by N_1 in 3.3 so that (14) is less sharp than the inequality $C \geq \bar{V} = N_3$.

By appropriate modification of the foregoing (see 2.7 A (16)) we obtain the inequality

$$(15)\qquad C \geq \frac{1+\nu_1}{4\pi\nu_1} \cdot \iint_{\underline{A}} r d\omega = \left(\frac{S'}{S}\right)^{1/2} \iint \frac{\cos^2(r,n)\, d\sigma}{4\pi d}$$

for the capacity of a condenser bounded by the similar surfaces $\underline{A} = \underline{A}_0$ and $\underline{A}' = \underline{A}_{\nu_1}$.*

(e) We can deal in a similar fashion with the capacity c per unit height of a cylindrical condenser whose cross-sections are certain curves $\underline{C} = \underline{C}_0$ and $\underline{C}' = \underline{C}_{\nu_1}$ similar and similarly situated with respect to a point p_0 in the interior of $\underline{A}$. Let h be the distance of a tangent of $\underline{C}$ from p_0. Then, by Dirichlet's principle, the inequality

$$(16)\qquad c \leq \frac{1}{4\pi \log(1+\nu_1)} \int_{\underline{C}} \frac{ds}{h} = \frac{B_0}{4\pi \log(1+\nu_1)}$$

holds; this can be written also in the alternate forms:

$$(17)\qquad c \leq \frac{1}{4\pi \log(L'/L)} \int_{\underline{C}} \frac{ds}{h} = \frac{(L'/L) - 1}{\log(L'/L)} \int_{\underline{C}} \frac{ds}{4\pi d} .$$

---

* This inequality is again less sharp than the inequality resulting from the analogue of Poincaré's theorem for condensers (cf. 30, p. 584).

Here L and L' have the usual meaning for the given curves and d denotes the distance between corresponding parallel tangents to the two coatings. Thomson's principle yields:

$$(18) \qquad c \geqq \frac{1}{2 \log (1 + \nu_1)} = \frac{1}{2 \log (L'/L)} \quad .$$

We may compare directly the upper and lower bounds (16) and (18). We find by an argument similar to that in (b):

$$(19) \qquad L^2 \leqq 2A \int_C h^{-1} \, ds = 2 A B \, .$$

Hence by the isoperimetric inequality

$$(20) \qquad B \geqq 2 \pi \, .$$

3.6. Surfaces of revolution. (a) We consider a surface of revolution, represent the meridian curve in a complex w-plane and choose the real axis as axis of symmetry. We denote the outer radius of the meridian curve by $\bar{r}$. Let $w = f(z)$ be the mapping function of the exterior of this curve onto the exterior of the circle $|z| = \bar{r}$ such that

$$(1) \qquad w = f(z) = z + c_0 + c_1 z^{-1} + \ldots + c_n z^{-n} + \ldots \quad .$$

Then all coefficients c_n are real. Obviously $\Im f(re^{i\theta})$ is positive for $0 < \theta < \pi$. We seek relations between the capacity C of the given surface and the outer radius $\bar{r}$ of its meridian curve.

As a consequence of the principles stated in Chapter II we shall prove the inequalities:

$$(2) \qquad \int_0^{\pi} \frac{d\theta}{\displaystyle\int_{\bar{r}}^{\infty} \frac{dr}{r \, \Im f(re^{i\theta})}} \leqq 2C \leqq \frac{\pi}{\displaystyle\int_{\bar{r}}^{\infty} \frac{dr}{r \| \Im f \|}} \quad .$$

Here the following abbreviation is used:

$$(3) \qquad \| \Im f \| = \frac{1}{\pi} \int_0^{\pi} \Im f(re^{i\theta}) \, d\theta \quad .$$

(b) Before proving (2), we rewrite it in the form

$$(4)\qquad \int_0^{\pi}\left\{\int_{\bar r}^{\infty}[u(r,\theta)]^{-1}\,dr\right\}^{-1}d\theta \leqq C \leqq \left\{\int_{\bar r}^{\infty}\left[\int_0^{\pi}u(r,\theta)d\theta\right]^{-1}dr\right\}^{-1};$$

we have used the abbreviation

$$(5)\qquad r\,\Im f(re^{i\theta})\,/\,2 = u(r,\theta)\ .$$

That the upper bound in (4) is not less than the lower bound, can be shown independently of the particular nature of u or the particular values of the limits of integration.* The two bounds coincide, if and only if,

$$u(r,\theta) = g(r)\,h(\theta)$$

and this can happen if, and only if,

$$f(z) = z \pm \rho^2/\,z\ ,\qquad \rho \geqq 0,\quad \bar r = 1\ ,$$

except for trivial transformations. Indeed, if $\Im f(re^{i\theta})$ has to be a product of two functions, each of one variable only, then

$$\Im f(re^{i\theta}) = (Ar^n + Br^{-n})\sin(n\,\theta + \gamma)$$

where n, A, B and γ are constants, n an integer; this follows in the usual way from Laplace's equation in polar coordinates. Yet f(z) should be univalent, and this leaves only $n = \pm 1$.

The two forms of f(z), which we have thus obtained, correspond to the two kinds of spheroids, prolate and oblate. That is, <u>for spheroids, and only for spheroids, both inequalities</u> (2) <u>become equations.</u>

In fact, in the following proof of (2) we shall choose the level surfaces and the lines of force so that they coincide with the true surfaces and lines in the case of the spheroids.

(c) <u>Proof of the upper bound</u>. Applying 2.5 (9) we choose for the level surface $\underline{A}(\nu)$ the surface of revolution arising by rotation of the level curve in the w-plane which corresponds to the circle $|z| = r$, $r > \bar r$, in the conformal mapping $w = f(z)$. We denote by ds a line-element of the level curve, and by φ the angle between the plane of this curve and a fixed plane. Then, following the notation in 2.5:

$$\nu = r/\bar r - 1,\qquad d\nu = dr/\bar r\ ,$$

$$dn = |f'(re^{i\theta})|\,dr\ ,$$

* See G. H. Hardy, J. E. Littlewood and G. Pólya, Inequalities, p. 148, th. 202, to which the present case is analogous, although not contained in it ($k = -1$).

$$d\sigma = ds \cdot \mathfrak{J} f(re^{i\theta}) d\varphi = |f'(re^{i\theta})| \; \mathfrak{J} f(re^{i\theta}) \; r d\theta \, d\varphi$$

so that

$$(6) \qquad T(\nu) = \frac{1}{4\pi} \int_{\theta=0}^{\pi} \int_{\varphi=-\pi}^{\pi} (r/\bar{r}) \; \mathfrak{J} f(re^{i\theta}) \, d\theta \, d\varphi = \frac{\pi r}{2 \bar{r}} \, \|\mathfrak{J} f\| .$$

This yields the upper bound in (2) immediately.

Proof of the lower bound. We apply the principle of Thomson in the form 2.7 (6). For the lines of force we choose plane curves corresponding in the meridian plane to the radii θ = const.. We have, following the notation in 2.7,

$$\lambda = 0, \qquad dl = |f'(re^{i\theta})| \, dr,$$

$$dS = |f'(re^{i\theta})| \; \mathfrak{J} f(re^{i\theta}) \; r d\theta \, d\varphi,$$

$$d\sigma = |f'(\bar{r}e^{i\theta})| \; \mathfrak{J} f(\bar{r}e^{i\theta}) \; \bar{r} d\theta \, d\varphi$$

so that

$$(7) \qquad L(p) = \frac{4\pi \; d\sigma}{d\theta \;\; d\varphi} \int_{\bar{r}}^{\infty} \frac{dr}{r \, \mathfrak{J} f(re^{i\theta})} .$$

This furnishes the lower bound in (2) immediately.

3.7. An inequality between C and $\bar{r}$. The foregoing discussion suggests that the maximum of the ratio $C/\bar{r}$ might be attained for the limiting case $\rho = 1$ of the oblate spheroid, that is, for a circular disk; see 3.6 (b). This is equivalent to the inequality

$$(1) \qquad C \leq \frac{4}{\pi} \bar{r} .$$

(There is obviously no minimum for the ratio $C/\bar{r}$ since it will be arbitrarily small for a prolate spheroid if ρ is sufficiently near 1; this is the case of the "needle.")

(a) From 3.6 (1) we obtain $\mathfrak{J} f(re^{i\theta}) = r \sin \theta - \sum_{n=1}^{\infty} c_n r^{-n} \sin n\theta$ and so from 3.6 (3)

$$(2) \qquad \pi \|\mathfrak{J} f\| = 2r - 2 \sum_{n \text{ odd}, \; n \geq 1} c_n r^{-n}/n$$

Therefore, the second inequality 3.6 (2) becomes

$$(3) \qquad \frac{1}{C} \geq \int_{\bar{r}}^{\infty} \frac{dr}{r^2 - c_1 - \frac{c_3}{3r^2} - \frac{c_5}{5r^4} - \frac{c_7}{7r^6} - \dots}$$

We assume now that

$$(4) \qquad \bar{r} = 1 .$$

Inequality (1) follows then if we can prove that

$$(5) \qquad c_1 + \frac{c_3}{3r^2} + \frac{c_5}{5r^4} + \frac{c_7}{7r^6} + \dots \geq -1, \qquad r \geq 1 .$$

This is equivalent with the estimate

$$(6) \qquad \| \mathfrak{J} f \| \leq \frac{2}{\pi} (r + r^{-1}) , \qquad r \geq 1 .$$

(b) We proved (5) first under the restriction that $\Re\, f(e^{i\theta})$ is a decreasing function of θ in the interval $0 < \theta < \pi$.* Another formulation of this condition is that each plane, perpendicular to the axis of revolution and intersecting the given solid, intersects it in a circular disk. Yet, in fact, no such restriction is needed.

[Inequality (5) can be written in the form

$$\int_r^{\infty} \left(\frac{c_1}{\rho^2} + \frac{c_3}{\rho^4} + \frac{c_5}{\rho^6} + \dots \right) d\rho \geq -\frac{1}{r}, \quad \text{or}$$

$$(7) \qquad \int_r^{\infty} \left(\frac{f(\rho) - f(-\rho)}{2\rho} - 1 + \frac{1}{\rho^2} \right) d\rho \geq 0 , \qquad r \geq 1 ,$$

and this follows quite generally from the following theorem of Golusin: Let $f(z) = z + c_0 + c_1 z^{-1} + c_2 z^{-2} + \dots$ be regular and schlicht for $|z| > 1$; then

$$(8) \qquad \left| \frac{f(z') - f(z'')}{z' - z''} \right| \geq 1 - \frac{1}{r^2}$$

provided $|z'| = |z''| = r$, $r > 1$.** We choose $z' = -z'' = \rho$; in our case

* Concerning mappings of this kind, cf. L. Fejér, Neue Eigenschaften der Mittelwerte bei den Fourierreihen, Journal of the London Math. Society, vol.8 (1933) pp. 53-62; p. 61.

** G. Golusin, On distortion theorems and coefficients of univalent functions, Recueil Mathématique (Math. Sbornik) N.S. 19 (61) (1946) pp. 183-202. (Russian, English summary.)

$f(\rho) - f(-\rho)$ is real and positive. Hence we have proved (5), (6) and (1) generally.]

(c) Another upper bound for the capacity C can be obtained as follows. We define the positive function

$$g(r) = 1 - \frac{c_1}{r^2} - \frac{c_3}{3r^4} - \frac{c_5}{5r^6} - \dots \quad . \tag{9}$$

By Schwarz's inequality

$$\int_1^\infty r^{-2} [g(r)]^{-1} \, dr \int_1^\infty r^{-2} g(r) \, dr \geqq [\int_1^\infty r^{-2} \, dr]^2 = 1 \tag{10}$$

so that from (3)

$$C \leqq \int_1^\infty r^{-2} g(r) \, dr ;$$

in view of (9)

$$C \leqq 1 - \frac{c_1}{1.3} - \frac{c_3}{3.5} - \frac{c_5}{5.7} - \frac{c_7}{7.9} - \dots \tag{11}$$

follows. This estimate will be soon useful.

(d) We assume now that the curve $\underline{C}$ in the w-plane, the exterior of which is mapped onto $|z| > \bar{r}$ by 3.6 (1), is symmetrical with respect to both coordinate axes. Then 3.6 (1) must be an odd function,

$$c_{2n} = 0 \qquad \text{for} \quad n = 0, 1, 2, \dots \quad . \tag{12}$$

Rotating $\underline{C}$ about the real axis, we obtain the surface of revolution with capacity C which we have discussed heretofore. Rotating $\underline{C}$ about the imaginary axis, we obtain another surface of revolution with capacity C'. The transition from C to C' means substituting

$$w = -i\, f(iz) = z - c_1 z^{-1} + c_3 z^{-3} - c_5 z^{-5} + \dots$$

for 3.6 (1); we took (12) into account. Therefore, supposing (4), we conclude from (11) that

$$C' \leqq 1 + \frac{c_1}{1.3} - \frac{c_3}{3.5} + \frac{c_5}{5.7} - \frac{c_7}{7.9} + \dots$$

and hence

$$C + C' \leqq 2 - 2\left(\frac{c_3}{3.5} + \frac{c_7}{7.9} + \frac{c_{11}}{11.13} + \dots\right) \tag{13}$$

Using the area theorem

(14) $$c_1^2 + 2c_2^2 + 3c_3^2 + \dots \leqq 1$$

and Cauchy's inequality, we obtain

(15) $$C + C' \leqq 2 + 2\left[\sum_{n=1}^{\infty} (4n-1)^{-3}(4n+1)^{-2}\right]^{1/2}$$

This yields, if we lift the restriction (4)

(16) $$C + C' < 2.08\,\bar{r} .$$

In the particular case when the curve $\underline{C}$ is an ellipse, and both solids generated by its rotation are spheroids, the one oblate, the other prolate, $c_3 = c_7 = \dots = 0$; see 3.6 (b). Therefore, in this case, we find from (13) that

$$C + C' \leqq 2\,\bar{r} ;$$

the sign of equality is attained in the case of the sphere.

3.8. On the capacity of a cube. (a) There is some evidence that Dirichlet was in possession of a representation for the potential of the electrostatic equilibrium distribution on the surface of a cube.* The case of the cube may serve as an interesting illustration of the general methods for obtaining upper and lower bounds for the capacity of a solid.

(b) The following estimates hold for the capacity C of a cube with edge a :

(1) $$0.632\,a < C < 0.71055\,a .$$

In what follows we assume that $a = 1$. Comparing first our cube with the inscribed and circumscribed spheres we obtain easily $C > 0.5$ and $C < 3^{1/2}/2 = 0.865$.

A much better upper bound can be obtained by using Minkowski's constant M. Each edge contributes to M the half of its length multiplied by the angle of the exterior normals along this edge. Hence

$$M = 12 \cdot \frac{1}{2} \cdot \frac{\pi}{2} = 3\pi , \qquad \bar{M} = 3/4 ,$$

* Cf. G. L. Dirichlet, Werke, vol. 2, p. 420 (remark by H. Weber). - Cf. also, L. Koenigsberger, C. G. J. Jacobi, p. 363 (letter of Jacobi to F. Neumann).

so that $C < 0.75$. Finally the inequality 3.4 (4) involving S and M furnishes the upper bound in (1). We remind the reader that this estimate is a consequence of Dirichlet's principle applied in the form 2.5 (9). In this case the level surfaces used are the parallel surfaces of the cube; these consist, as in the case of any polyhedron, of pieces of planes, cylinders and spheres.

(c) A very good lower bound follows from the inequality

$$C > \bar{V} = (3/(4\pi))^{1/3} = 0.62033 \quad . \tag{2}$$

Better lower bounds can be obtained by means of Gauss-Thomson's principle in the form given in 2.9 (b). We observe, in particular, the remark at the end of 2.9 (b). If the origin of our coordinate system x,y,z is at the center of the cube, the only harmonic polynomials to be considered will be those having the following two properties:

(1) they are symmetrical in x,y,z;

(2) they contain only even powers of x,y,z.

By a very simple discussion we ascertain that up to the degree 8 the only homogeneous polynomials of this kind are (except for constant factors):*

$$1,$$

$$x^4 + y^4 + z^4 - 3(y^2z^2 + z^2x^2 + x^2y^2),$$

$$x^6 + y^6 + z^6 - \tfrac{15}{2}\,(x^4(y^2 + z^2) + y^4(z^2 + x^2) + z^4(x^2 + y^2)) + 90\, x^2y^2z^2,$$

$$x^8 + y^8 + z^8 - 14\,(x^6(y^2 + z^2) + y^6(z^2 + x^2) + z^6(x^2 + y^2)) + 35\,(y^4z^4 + z^4x^4 + x^4y^4) \;.$$

The four first minima corresponding to these functions (furnishing upper bounds for $1/C$) have been computed. They give the following lower bounds for C:

---

* It can be shown that the only terms which can actually occur in the expansion of the potential of the cube, must have the form

$$\text{const.}\; D_1^m D_2^n \,(1/r)\,,$$

where D_1 and D_2 denote the following two differential operators:

$$D_1 = \frac{\partial^2}{\partial x^2}\frac{\partial^2}{\partial y^2}\frac{\partial^2}{\partial z^2}\,,$$

$$D_2 = \left(\frac{\partial}{\partial x} + \frac{\partial}{\partial y} + \frac{\partial}{\partial z}\right)\left(-\frac{\partial}{\partial x} + \frac{\partial}{\partial y} + \frac{\partial}{\partial z}\right)\left(\frac{\partial}{\partial x} - \frac{\partial}{\partial y} + \frac{\partial}{\partial z}\right)\left(\frac{\partial}{\partial x} + \frac{\partial}{\partial y} - \frac{\partial}{\partial z}\right) \;.$$

The degree of the corresponding harmonic polynomial is $6m + 4n$. [The spherical harmonics with any given crystallographic symmetry can be similarly determined. See G. Pólya and B. Meyer, Sur les symétries des fonctions sphériques de Laplace, Comptes Rendus de l'Académie des Sciences, vol. 228 (1949) pp. 28-30; Sur les fonctions sphériques de Laplace de symétrie cristallographique donnée, ibidem, pp. 1083-1084.]

0.601547,
0.622948,
0.62875,
0.632.

The first bound is identical with $N_{-\Pi}$ in the notation of 3.3. As it was pointed out there, this bound is never as good as $\overline{V}$. In our case the next bound is already better than $\overline{V}$.

The necessary computations, in particular in the fourth step, are rather involved.

[Prof. T. J. Higgins wrote us on Dec. 8, 1949 that he obtained $C \sim 0.65$ and Prof. M. Picone informed us on Feb. 23, 1950 that W. Gross, working in his institute, obtained $C \sim 0.646$ and communicated this result to the Innsbruck meeting of the Austrian Math. Society in Sept. 1949. In both cases the numerical work seemed to indicate that the value obtained is a fairly close lower approximation.*]

[3.9. On the capacity of a square. In the present section C stands for the capacity of a square plate with side a. We mention an approximate value, a lower and an upper bound:

(1) $C \sim 0.3607a$

(2) $C > 2\pi^{-3/2}a = 0.35917a$

(3) $C < 2^{-1}\pi^{-5/2}[\Gamma(1/4)]^2 = 0.37572a$.

The approximate value (1) is due to Maxwell**; his method, however, gives no information about the sign or amount of the deviation of the approximate value from the true value. The lower bound (2) is due to Lord Rayleigh (26, vol. 2, pp. 176-180). When stated, it was based on an unproved conjecture asserting that of all plates with a given area A the circular plate has minimum capacity; yet this conjecture has been proved since (24; see 7.3(d)). The upper bound is based on an unproved conjecture asserting that of all plates with a given outer radius $\bar{r}$ the circular plate has maximum capacity (24; entry 17 in the table of sect. 1.21). None of the upper bounds for C that we can completely justify comes close to (3). The method of sect. 2.9, applied in the foregoing sect. 3.8 to the cube, does not apply to the square or to any other plate.

A statement of Maxwell (loc.cit.) deserves to be quoted: "the capacity of a rectangle . . . is greater than that of a square of equal area." This is true (with $\geq$, see 7.4(b)), yet Maxwell's proof is amazingly fallacious.]

* D. K. Reitan and T. J. Higgins, Calculation of the electrical capacitance of a cube, Jour. Appl. Phys., vol. 22 (1951) pp. 223-226.

** H. Cavendish, Electrical researches, edited by J. C. Maxwell, 1879. See pp. 426-427.

CHAPTER IV. CIRCULAR PLATE CONDENSER

4.1. Problem. A circular plate condenser consists of two congruent circular disks $\underline{A}_0$ and $\underline{A}_1$ with a common axis. We denote the radii of these disks by a, the distance between their planes by c. We charge them with electricity of equal amount but opposite sign denoted by $\pm$ Q; let $Q > 0$ and let Q and -Q be the charges of $\underline{A}_0$ and $\underline{A}_1$, respectively. In the position of equilibrium the potential will be constant on the disks and equal to $\pm V_0$. The capacity $C = Q/(2V_0)$ of the condenser was defined in 2.2 (d).

The most interesting problem related to C is to find approximations of C for small values of the ratio $c/a = q$. Such approximations have been proposed by Kirchhoff (13) and Ignatowsky (12). Kirchhoff obtained

$$(1) \qquad C \sim \frac{a}{4q} + \frac{a}{4\pi} \log \frac{1}{q} + \frac{a}{4\pi}\left(\log (16\pi) - 1\right)$$

whereas Ignatowsky found

$$(2) \qquad C \sim \frac{a}{4q} + \frac{a}{4\pi} \log \frac{1}{q} + \frac{a}{4\pi}\left(\log 8 - \frac{1}{2}\right) .$$

The arguments of these authors do not seem to be quite complete. Also the calculation leading to (2) is extremely long and complicated.

In this Chapter we prove, by using Gauss' principle (2.8), that

$$(3) \qquad \frac{C}{a} > \frac{1}{4q} + \frac{1}{4\pi} \log \frac{1}{q} + \frac{1}{4\pi}\left(\log 8 - \frac{1}{2}\right) + \varepsilon$$

where $\lim \varepsilon = 0$ as $q \to 0$.

4.2. Lower bound for the capacity of a circular plate condenser.
(a) We apply the principle of Gauss assuming uniform charges of equal magnitude and opposite sign over the disks $\underline{A}_0$ and $\underline{A}_1$. These charges approximate reasonably well the true electrostatic equilibrium distribution which is known to be practically uniform except around the edges. We choose $Q = 1$ so that the constant density function on $\underline{A}_0$ is

(1) $$\mu = (a^2\pi)^{-1}$$

and the opposite on $\underline{A}_1$. The potential u of the total charge is also opposite at corresponding points of $\underline{A}_0$ and $\underline{A}_1$ so that by 2.8 (5)

(2) $$\frac{1}{C} \leqq \frac{2}{a^2\pi} \iint_{\underline{A}_0} u\, d\sigma .$$

(b) Introducing cylindrical coordinates x, ρ we have, as is well known,* for the potential of a uniformly charged disc $\underline{A}_0$, the total charge being 1,

(3) $$\int_0^\infty e^{-xt} J_0(\rho t)\, 2\, \frac{J_1(at)}{at}\, dt ;$$

here $x > 0$ is the distance from the plane of $\underline{A}_0$ and ρ is the distance from the axis. Consequently we have on $\underline{A}_0$

(4) $$u = u(\rho) = \int_0^\infty (1 - e^{-ct})\, J_0(\rho t)\, 2\, \frac{J_1(at)}{at}\, dt , \qquad 0 \leqq \rho \leqq a,$$

so that, by (2),

(5) $$\frac{1}{C} \leqq \frac{2}{a^2\pi} \int_0^a \rho u(\rho)\, d\rho \cdot 2\pi = \frac{4}{a^2} \int_0^a \rho u(\rho)\, d\rho .$$

But

(6) $$\int_0^a \rho J_0(\rho t) d\rho = \frac{a^2}{2}\, 2\, \frac{J_1(at)}{at} ;$$

hence

$$\frac{1}{C} \leqq \frac{4}{a^2} \cdot \frac{a^2}{2} \int_0^\infty (1 - e^{-ct}) \left(2\, \frac{J_1(at)}{at}\right)^2 dt$$

(7) $$= \frac{8}{a} \int_0^\infty (1 - e^{-qt}) \left(\frac{J_1(t)}{t}\right)^2 dt .$$

---

*. Cf. A. G. Webster - G. Szegö, Partielle Differentialgleichungen der Mathematischen Physik, 1930, p. 441 (242).

(c) The function

$$(8) \qquad \Phi(q) = \int_0^\infty e^{-qt} \left(\frac{J_1(t)}{t} \right)^2 dt$$

can be expressed in terms of the complete elliptic integrals K and E
associated with the modulus

$$(9) \qquad k = (1 + q^2/4)^{-1/2} \; .$$

We have [*]

$$(10) \qquad \Phi(q) = -\frac{q}{2} + \frac{4(1-k^2)}{3\pi k^3} K + \frac{4(2k^2-1)}{3\pi k^3} E$$

so that using the customary approximations [**] of K and E as $q \to 0$, $k \to 1$,

$$(11) \qquad \Phi(q) = -\frac{q}{2} + \frac{4}{3\pi} - \frac{q^2}{2\pi} \log \frac{q}{8} - \frac{q^2}{4\pi} + O\left(q^4 \log \frac{1}{q}\right) \; .$$

Consequently

$$\frac{1}{C} \leq \frac{8}{a}(\Phi(0) - \Phi(q))$$

$$(12) \qquad = \frac{4q}{a} \left(1 + \frac{q}{\pi} \log \frac{q}{8} + \frac{q}{2\pi} \right) + O\left(q^4 \log \frac{1}{q}\right)$$

from which 4.1 (3) follows.

4.3. <u>Generalization</u>. It is of some interest to generalize the previous argument. This generalization leads not only to another upper bound of 1/C but, at least theoretically, to an exact formula for 1/C itself.

(a) We consider on $\underline{A}_0$ an arbitrary circular-symmetrical distribution defined by the density function

$$(1) \qquad \mu(\rho) = \frac{1}{a^2\pi} \sum_{m=0}^\infty x_m P_m \left(1 - 2\frac{\rho^2}{a^2}\right), \qquad 0 \leq \rho \leq a \; ,$$

where $x_0 = 1$. Here P_m denotes the Legendre polynomials, so that the total charge on $\underline{A}_0$ is

$$(2) \qquad Q = 2\pi \int_0^a \mu(\rho) \, \rho \, d\rho = x_0 = 1 \; .$$

---

[*] Lord Rayleigh, Scientific Papers, vol. 4, 1904, p. 260; G. N. Watson, A treatise on the theory of Bessel functions, p. 389; cf. Ignatowsky <u>12</u>, p. 30.
[**] E. Jahnke-F. Emde, Funktionentafeln, Tables of functions, 4. edition, 1945, p. 73.

We assume on $\underline{A}_1$ the opposite charge and denote by u the potential of this whole distribution; we have* on $\underline{A}_0$

$$u = u(\rho) = \int_0^\infty (1 - e^{-ct}) J_0(\rho t) h(t) \, dt \tag{3}$$

where

$$h(t) = 2\pi \int_0^a \rho\mu(\rho) J_0(\rho t) \, d\rho . \tag{4}$$

Hence by 2.8 (5)

$$\frac{1}{C} \leq 2 \int_0^a \rho u(\rho) \mu(\rho) \, d\rho \cdot 2\pi$$

$$= 4\pi \int_0^a \rho\mu(\rho) u(\rho) \, d\rho . \tag{5}$$

Inserting here $u(\rho)$ and taking (4) into account we find

$$\frac{1}{C} \leq 4\pi \int_0^\infty (1 - e^{-ct}) \frac{h(t)}{2\pi} h(t) \, dt$$

$$= 2 \int_0^\infty (1 - e^{-ct}) [h(t)]^2 \, dt . \tag{6}$$

(b) An alternate form of this result can be obtained by using**

$$\int_0^a \frac{2\rho}{a^2} P_m \left(1 - 2\frac{\rho^2}{a^2}\right) J_0(\rho t) d\rho = 2 \frac{J_{2m+1}(at)}{at} , \qquad m = 0, 1, 2, \ldots; \tag{7}$$

we find

$$h(t) = 2 \sum_{m=0}^\infty x_m \frac{J_{2m+1}(at)}{at} , \tag{8}$$

so that $h(0) = x_0 = 1$. Thus

$$\frac{1}{2C} \leq \int_0^\infty (1 - e^{-ct}) \left(\sum_{m=0}^\infty 2x_m \frac{J_{2m+1}(at)}{at} \right)^2 dt, \qquad x_0 = 1 . \tag{9}$$

In fact $(2C)^{-1}$ is the minimum of the right-hand expression since formula (1) represents the most general density function on $\underline{A}_0$.

* Cf. Webster-Szegö, loc. cit., p. 433 (194), p. 434 (203).

** This follows from Watson, loc. cit., p. 404 (6), by applying Hankel's inversion formula.

(c) We can write

$$(10) \qquad \mu_n = \min \int_0^\infty (1 - e^{-ct}) \left(\sum_{m=0}^{n} 2x_m \frac{J_{2m+1}(at)}{at} \right)^2 dt \, , \qquad x_0 = 1 \; ;$$

the minimum is taken for all real values of x_m, $x_0 = 1$. Hence

$$(11) \qquad \mu_0 \geqq \mu_1 \geqq \mu_2 \geqq \cdots \geqq \mu_n \geqq \cdots \geqq (2C)^{-1}$$

and

$$(12) \qquad (2C)^{-1} = \lim_{n \to \infty} \mu_n \, .$$

It is not difficult to compute μ_n. Let us write

$$(13) \qquad 4 \int_0^\infty (1 - e^{-ct}) \frac{J_{2p+1}(at)}{at} \frac{J_{2p'+1}(at)}{at} \, dt = \lambda_{pp'}$$

Then $\mu_0 = \lambda_{00}$ and

$$(14) \qquad \mu_n = [\lambda_{pp'}]_0^n \Big/ [\lambda_{pp'}]_1^n \, , \qquad n = 1, 2, 3, \ldots \, .$$

Consequently

$$(15) \qquad (2C)^{-1} = \lim_{n \to \infty} [\lambda_{pp'}]_0^n \Big/ [\lambda_{pp'}]_1^n \, .$$

4.4. Another characterization of the capacity. (a) The constants $\lambda_{pp'}$ introduced in the previous section can be represented in another form. We start out from the following formula*

$$(1) \qquad \int_0^\infty e^{-ut} J_\nu(at) \frac{dt}{t^2} = \frac{1}{\nu} \int_u^\infty \left(\frac{(\alpha^2 + a^2)^{1/2} - \alpha}{a} \right)^\nu d\alpha \, , \qquad u > 0 \, ,$$

so that

$$(2) \quad \int_0^\infty (e^{-ut} - e^{-vt}) J_\nu(at) \frac{dt}{t^2} = \frac{1}{\nu} \int_u^v \left(\frac{(\alpha^2 + a^2)^{1/2} - \alpha}{a} \right)^\nu d\alpha \, , \quad 0 < u < v \, .$$

In (1) we must have $\nu > 1$, in (2) we must have $\nu > 0$. But**

$$(3) \quad J_{2p+1}(at) \, J_{2p'+1}(at) = \frac{2}{\pi} \int_0^{\pi/2} J_{2p+2p'+2}(2at \cos \theta) \cos (2p - 2p')\theta \, d\theta$$

---

* Cf. Watson, loc. cit., p. 386 (7).

** Watson, loc. cit., p. 390.

so that

$$\lambda_{pp'} = \frac{4}{a^2}\frac{2}{\pi}\int_0^{\pi/2}\cos(2p-2p')\theta \int_0^{\infty}(1 - e^{-ct})\, J_{2p+2p'+2}(2at\cos\theta)\frac{dt}{t^2}\,d\theta$$

(4)

$$= \frac{8}{a^2\pi}\,\frac{1}{2p+2p'+2}\int_{\theta=0}^{\pi/2}\int_{\alpha=0}^{c}\cos(2p-2p')\theta\left(\frac{(\alpha^2+4a^2\cos^2\theta)^{1/2}-\alpha}{2a\cos\theta}\right)^{2p+2p'+2} d\theta\, d\alpha$$

We introduce the notation

$$r = r(\alpha;\theta) = \frac{(\alpha^2 + 4\cos^2\theta)^{1/2} - \alpha}{2\cos\theta} \tag{5}$$

so that

$$\lambda_{pp'} = \frac{4}{a\pi}\,\frac{1}{p+p'+1}\int_{\theta=0}^{\pi/2}\int_{\alpha=0}^{q}\cos(2p-2p')\theta\,(r(\alpha;\theta))^{2p+2p'+2}\,d\theta\,d\alpha. \tag{6}$$

We put

$$\sum_{p=0}^{\infty} x_p r^{2p} e^{2pi\theta} = \sum_{p=0}^{\infty} x_p z^{2p} = f(z)\,, \qquad z = re^{i\theta}\,; \tag{7}$$

then

$$\sum_{p,p'=0,1,2,\ldots}\lambda_{pp'}x_p x_{p'} = \frac{8}{a\pi}\int_{\theta=0}^{\pi/2}\int_{\alpha=0}^{q}\int_{r=0}^{r(\alpha;\theta)}\left(\sum_{p,p'} r^{2p+2p'}e^{i(2p-2p')\theta}x_p x_{p'}\right) r\,dr\,d\theta\,d\alpha$$

$$= \frac{8}{a\pi}\int_0^q\left\{\iint_{M_\alpha}|f(z)|^2\,d\sigma\right\}d\alpha \tag{8}$$

where M_α is the domain $r \leq r(\alpha;\theta)$ in the first quadrant $0 \leq \theta \leq \pi/2$; $d\sigma$ is the area element.

Hence $(2C)^{-1}$ is the minimum of the expression (8) for all analytic functions $f(z)$ regular in the circle $|z| < 1$ which are real, even, and satisfy the condition $f(0) = 1$.

(b) The domain M_α is a part of the unit circle $|z| < 1$ which can be characterized in the following manner. The conformal mapping

$$w = z^{-1} - z \tag{9}$$

carries the domain M_α of the z-plane into the quarter plane $\Re w \geq \alpha$, $\Im w \leq 0$ of the w-plane.

Indeed, if $z = r\,e^{i\theta} = x + iy$, $w = u + iv$, we can write (5) as follows:

$$(2r \cos \theta + \alpha)^2 = \alpha^2 + 4 \cos^2 \theta ,$$

$$(2x + \alpha)^2 = \alpha^2 + \frac{4x^2}{x^2 + y^2} ,$$

$$x + \alpha = \frac{x}{x^2 + y^2} ,$$

(10) $$\alpha = u .$$

We have

(11) $$d\sigma = dx\,dy = \left|\frac{dz}{dw}\right|^2 du\,dv = \frac{du\,dv}{|1 + z^{-2}|^2} .$$

Let

(12) $$\frac{z^2 f(z)}{1 + z^2} = F(w) ;$$

this function $F(w)$ can be characterized by the following conditions:

I. it is regular in the complex w-plane cut along the segment $-2i$, $2i$ (the point $w = \infty$ inclusively);

II. about $w = \infty$ it is of the form

(13) $$F(w) = w^{-2} + c_4 w^{-4} + c_6 w^{-6} + \ldots$$

where c_4, c_6, ... are real constants.

For instance to the function $f(z) = 1$ the following function corresponds:

$$F(w) = \frac{1}{2} - \frac{1}{2}(1 + 4w^{-2})^{-1/2}$$

We write

(14) $$\Psi(u) = \int_{-\infty}^{\infty} |F(u + iv)|^2 \, dv , \qquad u > 0 .$$

Then we obtain the following further characterization of the capacity:

<u>The quantity</u> $(2C)^{-1}$ <u>is the minimum of the expression</u>

(15) $$\frac{4}{a\pi} \int_0^q \left\{ \int_\alpha^\infty \Psi(u)\, du \right\} d\alpha$$

<u>for all functions</u> $F(w)$ <u>satisfying the conditions</u> I. <u>and</u> II. <u>formulated above</u>.

4.5. <u>Generalization</u>. The following generalization of a circular plate condenser might be of interest. We consider two solids arising from each other by reflection with respect to a plane s. Charging the one solid with charge +Q, the other with charge -Q, the potential u of the equilibrium distribution will be constant, $u = u_0$ on the first, and $u = -u_0$ on the second solid. Then

$$C = \frac{Q}{2u_0} \tag{1}$$

is the capacity of the condenser.

It seems to be difficult to make a general statement about the approximate value of C when the two solids are near to each other. However it is possible to make such a statement if the smallest distance h of the given solids is <u>large</u>. Let C' be the capacity of the first solid with respect to the infinitely large sphere. Then for large h,

$$C = \frac{C'}{2}\left(1 + \frac{C'}{h}\right) + O\left(\frac{1}{h^2}\right) . \tag{2}$$

Suppressing the factor 1/2 on the right-hand side, we obtain the capacity of the condenser consisting of one of the solids and the plane s.

For the proof we consider the free charge +Q on the first solid which is in equilibrium distribution (its potential is Q/C' on this solid) and the charge -Q on the second solid with the opposite distribution. Denoting by v the resulting total potential we apply 2.12 (4). We have $\Gamma = Q$ and on the first solid:

$$v = \frac{Q}{C'} - \frac{Q}{h} + O\left(\frac{1}{h^2}\right) .$$

On the second solid we have the same expression with the opposite sign. Hence

$$\frac{Q}{C} = 2\left(\frac{Q}{C'} - \frac{Q}{h}\right) + O\left(\frac{1}{h^2}\right) . \tag{3}$$

Hereupon (2) follows readily.

In the case of the circular plate condenser treated in the previous part of this Chapter, we have $C' = 2a/\pi$, $h = c$, so that for large c,*

$$C = \frac{a}{\pi}\left(1 + \frac{2a}{\pi c}\right) + O\left(\frac{1}{c^2}\right) . \tag{4}$$

* Cf. Ignatowsky <u>12</u>, p. 53 (10).

CHAPTER V. TORSIONAL RIGIDITY AND PRINCIPAL FREQUENCY

Consequences of the Variational Definition

5.1. Characterization by a minimum property. We consider a simply connected plane domain $\underline{D}$, its boundary curve $\underline{C}$, its torsional rigidity P and its principal frequency Λ. Both Λ and P can be defined mathematically in various ways. The definition that is most useful for our purposes is based on a minimum property.

Let $f = f(x,y)$ denote a function, defined and sufficiently "smooth" in $\underline{D}$ and on $\underline{C}$, satisfying the condition

$$(1) \qquad f = 0 \quad \text{on } \underline{C} ,$$

but arbitrary otherwise. Then (as stated already in 1.26 and 1.27)

$$(2) \qquad \frac{\iint (f_x^2 + f_y^2)\, dx\, dy}{4(\iint f\, dx\, dy)^2} \geq \frac{1}{P} ,$$

$$(3) \qquad \frac{\iint (f_x^2 + f_y^2)\, dx\, dy}{\iint f^2\, dx\, dy} \geq \Lambda^2 .$$

For the double integrals occurring in (2) and (3), the domain of integration is $\underline{D}$. We lay down here the convention for the entire present chapter that any double integral without specified limits is extended over $\underline{D}$, unless the contrary is explicitly stated.

We define P^{-1} as the greatest number for which (2) holds and Λ^2 is similarly defined by (3). As the case of equality can be attained both in (2) and in (3), P^{-1} and Λ^2 are determined as minima in analogous problems of the calculus of variations.

These "variational" definitions are very important. The proof of the theorem (1.10) that symmetrization diminishes both 1/P and Λ^2 is based on them (see note A) and so will be various estimates that we shall derive in the sections 5.4 - 5.8. Yet other definitions (also mentioned in the introductory chapter, 1.26 - 1.27) are more familiar to the physicist. Therefore, we shall begin by establishing the equivalence of the variational and the non-variational definitions first for P and then for Λ; see 5.2 and 5.3.

5.2. Equivalence of the two definitions for the torsional rigidity. We consider a function $v = v(x,y)$ (the stress function) that satisfies the differential equation

(1) $$v_{xx} + v_{yy} + 2 = 0$$

in the interior of the domain D and the boundary condition

(2) $$v = 0 \quad \text{on } C.$$

There is such a function v. In fact, put

(3) $$v = \Phi - \frac{1}{2}(x^2 + y^2) .$$

Conditions (1) and (2) are equivalent to demanding that the function Φ should be harmonic inside D and should take prescribed boundary values:

(1') $$\Phi_{xx} + \Phi_{yy} = 0 \quad \text{in } D ,$$

(2') $$\Phi = \frac{1}{2}(x^2 + y^2) \quad \text{on } C .$$

It is well known that such a function Φ exists and is uniquely determined.

It is shown in textbooks of the theory of elasticity* that

(4) $$P = 2 \iint v dx dy$$

We must prove that this definition of P is equivalent to that given in connection with 5.1 (2).

We consider, therefore, a function f satisfying 5.1 (1). By the differential equation (1)

$$2 \iint f \, dx \, dy = - \iint f(v_{xx} + v_{yy}) \, dx \, dy$$

$$= \iint (f_x v_x + f_y v_y) \, dx \, dy - \iint [\partial(fv_x)/\partial x + \partial(fv_y)/\partial y] \, dx \, dy .$$

It follows, therefore, from 5.1 (1) that

(5) $$2 \iint f \, dx \, dy = \iint (f_x v_x + f_y v_y) \, dx \, dy .$$

The function v, satisfying (2), is a particular function f. Applied to this case, (5) yields in conjunction with (4)

* See, for instance, I. S. Sokolnikoff, Mathematical Theory of Elasticity, 1946, p. 189.

(6) $$P = 2 \iint v \, dx \, dy = \iint (v_x^2 + v_y^2) \, dx \, dy .$$

Hence it follows that equality is attained in 5.1 (2) when $f = v$.

For a general f, it follows from (5) that

$$(2 \iint f \, dx \, dy)^2 \leqq \left(\iint (v_x^2 + v_y^2)^{1/2} (f_x^2 + f_y^2)^{1/2} \, dx \, dy \right)^2$$

(7) $$\leqq P \iint (f_x^2 + f_y^2) \, dx \, dy .$$

We used in succession Cauchy's inequality, Schwarz's inequality and the equation (6). Having proved (7), we have arrived at the desired inequality 5.1 (2).

If the case of equality is attained in both inequalities under (7), there exists a constant c such that

$$f_x^2 + f_y^2 = c^2(v_x^2 + v_y^2), \qquad f_x = c \, v_x, \qquad f_y = c \, v_y ,$$

and we conclude finally from (2) and 5.1 (1) that

$$f = cv .$$

Only for functions of this form is equality in 5.1 (2) attained.

5.3. <u>Equivalence of the two definitions for the principal frequency</u>. We consider now a function $w = w(x,y)$ that satisfies with a suitable non-negative real constant Λ a differential equation of the form

(1) $$w_{xx} + w_{yy} + \Lambda^2 w = 0$$

in the interior of the domain $\underline{D}$ and the boundary condition

(2) $$w = 0 \quad \text{on } \underline{C}.$$

Moreover we require

(3) $$w > 0 \quad \text{in } \underline{D}.$$

We take here for granted the existence of such a function w and of such a constant Λ. Our aim is to derive 5.1 (3). We proceed in three steps.

(a) We assume first that

(4) $$f = q \, w$$

where q is a smooth function defined in $\underline{D}$ and on $\underline{C}$. By assumption (4) and the differential equation (1)

$$(5) \qquad \iint [f_x^2 + f_y^2 - \Lambda^2 f^2]\, dx\, dy =$$

$$= \iint [(q_x w + q w_x)^2 + (q_y w + q w_y)^2 + q^2 w (w_{xx} + w_{yy})]\, dx\, dy$$

$$= \iint (q_x^2 + q_y^2)\, w^2\, dx\, dy + H$$

where

$$(6) \quad H = \iint [q^2(w_x^2 + w w_{xx}) + 2 q q_x\, w w_x + q^2 (w_y^2 + w w_{yy}) + 2 q q_y\, w w_y]\, dx\, dy$$

$$= \iint [\partial (q^2\, w w_x)/\partial x + \partial (q^2\, w w_y)/\partial y]\, dx\, dy$$

$$= 0$$

by the boundary condition (2). From (5) and (6) follows

$$(7) \quad \iint (f_x^2 + f_y^2)\, dx\, dy = \Lambda^2 \iint f^2\, dx\, dy + \iint (q_x^2 + q_y^2)\, w^2\, dx\, dy\ .$$

This proves 5.1 (3) and allows even discussing the case of equality. This cannot be attained unless q_x and q_y vanish identically and q is a constant. In this case, however, f is a constant multiple of w, by (4), and equality in 5.1 (3) is actually attained.

(b) We assume now that f vanishes not only on the boundary curve $\underline{C}$ but also in a full neighborhood of this curve so that the points in which f does not vanish form a set, completely in the interior of $\underline{D}$, on which w has a positive lower bound. Therefore, f/w = q is, as f itself, a smooth function and, herewith, the present case is reduced to the former case (a).

(c) An arbitrary function f, vanishing on the boundary $\underline{C}$, can be approximated by functions of the kind considered under (b), provided that $\underline{C}$ is somewhat smooth. E.g. if $\underline{C}$ is star-shaped with respect to a point p, we construct the curve $\underline{C}'$ contained in $\underline{C}$, similar to $\underline{C}$ and similarly situated, with p as center of similitude. We transplant the given function f from the interior of $\underline{C}$ onto the interior of $\underline{C}'$ by similarity and obtain f'. We define f' = 0 between $\underline{C}$ and $\underline{C}'$. For the partial derivatives of f' the curve $\underline{C}'$ may be a line of finite discontinuity, yet this leaves f' sufficiently smooth. At any rate f', or a slightly modified function, falls under the case (b). Letting $\underline{C}'$ tend to $\underline{C}$, we obtain 5.1 (3) generally, but without full discussion of the case of equality.

As we mentioned in 5.1, the definition of Λ^2 by 5.1 (3) implies the uniqueness of Λ^2. We have shown, therefore, that there cannot be more than one number Λ^2 for which all three conditions, (1), (2) and (3), are satisfied. This Λ^2 cannot be zero; otherwise, w would be harmonic and for a harmonic function (2) and (3) are incompatible. Therefore, Λ^2 is positive, and we choose Λ as positive.

5.4. <u>An inequality connecting</u> P, Λ <u>and</u> A. Applying 5.1 (2) and 5.1 (3) to the function v, considered in 5.2, we obtain

$$\frac{1}{P} = \frac{\iint (v_x^2 + v_y^2)\, dx\, dy}{4\left(\iint v\, dx\, dy\right)^2}$$

$$(1) \qquad > \frac{\iint (v_x^2 + v_y^2)\, dx\, dy}{4\, A \iint v^2\, dx\, dy}$$

$$\geq \frac{\Lambda^2}{4\, A} .$$

In fact, as we have proved in 5.2, the inequality 5.1 (2) goes over into an equation if v is substituted for f. In passing to the second line of (1), we applied Schwarz's inequality and observed that

$$\iint dx\, dy = A$$

is the area of $\underline{D}$; we were justified in excluding the equality, since v is certainly not a constant. Finally, we applied 5.1 (3) to f = v, which is permissible, since we have 5.2 (2). Thus, we have proved the inequality:

$$(2) \qquad P\, \Lambda^2 < 4\, A .$$

5.5. <u>Similar level lines</u>. <u>Estimate for</u> P. We assume now that $\underline{C}$ is star-shaped with respect to a point which we choose as the origin of a system of polar coordinates t, φ and let

$$(1) \qquad t = r(\varphi)$$

be the polar equation of $\underline{C}$.

We consider a function f of a variable point in $\underline{D}$ the level lines of which are similar to $\underline{C}$ and similarly situated, with the origin as center of similitude. Such a function has the form

$$(2) \qquad f = g[t/r(\varphi)]$$

where g stands for a smooth function of one variable. The function (2) should satisfy 5.1 (1). Therefore

$$(3) \qquad g(1) = 0 .$$

We find

$$(4) \qquad \iint f \, dx \, dy = \int_{-\pi}^{\pi}\int_{0}^{r} g(t/r)\, t \, dt \, d\varphi$$

$$= \int_{-\pi}^{\pi} r^2 \, d\varphi \int_0^1 g(\rho)\, \rho \, d\rho$$

$$= 2A \int_0^1 g(\rho)\rho \, d\rho$$

where A stands for the area of $\underline{D}$. Similarly

$$(5) \qquad \iint (f_x^2 + f_y^2)\, dx\, dy = \int_{-\pi}^{\pi}\int_0^r (f_t^2 + t^{-2} f_\varphi^2)\, t\, dt\, d\varphi$$

$$= \int_{-\pi}^{\pi}\int_0^r (r^{-2} + r'^2 r^{-4})\, [g'(t/r)]^2 \, t\, dt\, d\varphi$$

$$= \int_{-\pi}^{\pi} (1 + r'^2 r^{-2})\, d\varphi \int_0^1 [g'(\rho)]^2 \rho\, d\rho$$

$$= B \int_0^1 [g'(\rho)]^2 \rho\, d\rho$$

where B stands, as obvious abbreviation, for an integral which we shall transform. Let h denote the length of the perpendicular drawn from the origin to the tangent at a variable point of $\underline{C}$ and ds the line element at the same point. Then

$$ds^2 = (r^2 + r'^2)\, d\varphi^2 , \qquad h/r = r d\varphi/ds$$

and, therefore,

$$(1 + r'^2 r^{-2})\, d\varphi = (ds/d\varphi)^2 r^{-2} d\varphi = h^{-1} ds ,$$

$$(6) \qquad B = \int h^{-1} ds ;$$

the integral is extended over the closed curve $\underline{C}$.

Combining (4), (5) and 5.1 (2), we obtain

$$(7) \qquad \frac{1}{P} \leqq \frac{B}{16 A^2} \frac{\int_0^1 [g'(\rho)]^2 \rho\, d\rho}{\left(\int_0^1 g(\rho)\, \rho\, d\rho\right)^2}$$

It remains to choose $g(\rho)$ which must satisfy (3). Now, the level lines of (2) coincide with level lines of the true stress-function when $\underline{C}$ is an ellipse or, in particular, a circle and the origin coincides with the center. Accordingly, we choose $g(\rho)$ so that, in this particular case, (2) should turn out the true stress function. Therefore, we put

$$g(\rho) = \tfrac{1}{2}(1 - \rho^2) . \tag{8}$$

With this choice, (7) yields

$$P \geqq A^2 B^{-1} . \tag{9}$$

We add three remarks.

(a) The method of 2.5 and 2.7 shows that, choosing g according to (8), we obtain the minimum of the right hand side of (7). In fact, by (3) and Schwarz's inequality

$$\left(\int_0^1 g \rho \, d\rho\right)^2 = \left(- \int_0^1 g' \rho^{1/2} (\rho^{3/2}/2) \, d\rho\right)^2$$

$$\leqq \int_0^1 g'^2 \rho \, d\rho \cdot (1/16)$$

and equality is attained if g is the function (8) or a constant multiple of it.

(b) On the left hand side of (6) we should have written, more precisely, B_0 instead of B, in expressing by the subscript 0 that we have drawn the perpendicular h to a variable tangent of $\underline{C}$ from the origin. Generally, we call (as in 1.20 (b)) B_a the integral (6) if the perpendicular h is drawn from a point a inside $\underline{C}$. The estimate (9) should be actually written in the form

$$P \geqq A^2 B_a^{-1} \quad ;$$

it becomes the sharpest when a is chosen so inside $\underline{C}$ that B_a becomes a minimum. Let us call B this minimum of B_a. Then we can let stand (9) as expressing the sharpest estimate.

[(c) Inequality (9) can be extended to the case where the cross-section is ring-shaped, provided that its two boundary lines are similar and similarly located with the origin as center of similitude. Let A and A' denote the areas enclosed by the outer and inner boundary lines, respectively. (A' is the area of the hole, A > A'.) Then

$$B_0^{-1}(A^2 - A'^2) \leqq P \leqq (2\pi)^{-1} (A^2 - A'^2) . \tag{10}$$

The left hand inequality goes over into (9) when $A' = 0$. The right hand inequality is a special case of a theorem proved in 25.]

5.6. Similar level lines. Estimate for Λ. Let $\underline{C}$ be star-shaped, as in the foregoing section and assume again 5.5 (2). Then

$$\iint f^2 \,dx\,dy = 2A \int_0^1 [g(\rho)]^2 \rho \, d\rho \; ; \tag{1}$$

see 5.5 (4). Combining (1), 5.1 (3) and 5.5 (5) we obtain

$$\Lambda^2 \leqq \frac{B}{2A} \frac{\int_0^1 [g'(\rho)]^2 \rho \, d\rho}{\int_0^1 [g(\rho)]^2 \rho \, d\rho} \; . \tag{2}$$

Now, we choose $g(\rho)$ so that 5.5 (2) should turn out a representation of the true shape of the membrane, vibrating in its gravest mode, in the particular case when $\underline{C}$ is a circle. Therefore, we put

$$g(\rho) = J_0(j\rho) \; ; \tag{3}$$

$J_0(x)$ is the well-known Bessel function and j its first positive root, already mentioned in 1.3 (1); observe that 5.5 (3) is satisfied. With this choice, (2) yields

$$\Lambda^2 \leqq j^2 B(2A)^{-1} \; . \tag{4}$$

We take here B in the meaning defined in 5.5 (b). Our derivation of (4) supposes that we have chosen the origin from the outset in the most favorable manner; see 5.5 (b).

5.7. Conformal mapping. Estimate for P. (a) We consider now the one-to-one conformal mapping of the interior of the domain $\underline{D}$, situated in the z-plane, onto the interior of the unit circle in the ζ-plane. This mapping is represented by the power series

$$z = a + a_1\zeta + a_2\zeta^2 + \ldots + a_n\zeta^n + \ldots \tag{1}$$

convergent for $|\zeta| < 1$. The center of the unit circle, $\zeta = 0$, corresponds to the point $z = a$ in the domain $\underline{D}$, and $|a_1|$ is the inner radius of $\underline{D}$ with respect to a,

$$|a_1| = r_a \; . \tag{2}$$

We shall write (1) sometimes also in the abbreviated form

$$z = z(\zeta) \; . \tag{3}$$

We put

(4) $$z = x + iy\ , \qquad \zeta = \xi + i\eta = \rho e^{i\varphi}$$

with real x, y, ξ, η and φ and positive ρ ($\rho \geq 0$).

Let $f = f(z)$ be an arbitrary real-valued function of a variable point z of $\underline{D}$, vanishing on the boundary $\underline{C}$ of $\underline{D}$. We consider in particular such functions f the level lines of which are the images of the circles concentric with, and interior to, the unit circle of the ζ-plane. This is the case if, and only if; the level lines of the function $f[z(\zeta)]$ of ζ (we use the notation (3)) are just those concentric circles in the ζ-plane. That is, $f[z(\zeta)]$ should depend only on ρ, not on φ. Therefore

(5) $$f[z(\zeta)] = g(\rho)$$

where g is a suitable function of one variable. Since f(z) vanishes on $\underline{C}$,

(6) $$g(1) = 0\ .$$

If f is a function of the kind just considered

(7) $$\iint f\,dx\,dy = \iint_{|\zeta|<1} f\left|\frac{dz}{d\zeta}\right|^2 d\xi\,d\eta$$

$$= \int_{-\pi}^{\pi}\int_0^1 g(\rho)\left|\frac{dz}{d\zeta}\right|^2 \rho\,d\rho\,d\varphi$$

$$= \int_0^1 g(\rho)\int_{-\pi}^{\pi}\sum_{k=1}^{\infty}\sum_{l=1}^{\infty} ka_k\,l\bar{a}_l\,\zeta^{k-1}\,\bar{\zeta}^{l-1}\,\rho\,d\varphi\,d\rho$$

$$= 2\pi\int_0^1 g(\rho)\sum_{k=1}^{\infty} k^2\,|a_k|^2\,\rho^{2k-1}\,d\rho\ .$$

We used the notation introduced in (4) and (5), and familiar orthogonality properties.

Since $z = x + iy$ is an analytic function of $\zeta = \xi + i\eta$

(8) $$f_x^2 + f_y^2 = (f_\xi^2 + f_\eta^2)\left|\frac{d\zeta}{dz}\right|^2 ,$$

as is well known and easily verified. Therefore (Dirichlet's integral remains invariant under conformal mapping)

(9) $$\iint (f_x^2 + f_y^2)\,dx\,dy = \iint_{|\zeta|<1} (f_\xi^2 + f_\eta^2)\,d\xi\,d\eta$$

$$= 2\pi\int_0^1 [g'(\rho)]^2\,\rho\,d\rho\ .$$

Combining (7), (9) and 5.1 (2), we obtain

(10) $$\frac{1}{P} \leqq \frac{\int_0^1 [g'(\rho)]^2 \rho \, d\rho}{8\pi [\int_0^1 g(\rho) \sum_1^\infty k^2 |a_k|^2 \rho^{2k-1} \, d\rho]^2} .$$

(b) It remains to select $g(\rho)$, which must satisfy (6). Now, the level lines chosen coincide with the level lines of the true stress function when $\underline{C}$ is a circle. Accordingly, we select the same function 5.5 (8) as before, and for the same reasons. With this choice, (10) yields

(11) $$P \geqq 2\pi \left(\tfrac{1}{2} |a_1|^2 + \tfrac{2}{3} |a_2|^2 + \tfrac{3}{4} |a_3|^2 + \ldots\right)^2 .$$

Now let a be one of the points at which the inner radius attains its maximum. In symbols

$$\dot{r} = r_a = |a_1| ;$$

see (2). We conclude from (11) that

(12) $$P > \frac{\pi}{2} \dot{r}^4$$

unless $a_2 = a_3 = \ldots = 0$, that is, <u>unless D is a circle</u>. For a circle, (12) goes over into an equation.

[(c) We can however select $g(\rho)$ more advantageously; see <u>36</u>. We rewrite (10) in the form

(13) $$\frac{1}{P} \leqq \frac{\int_0^1 [g'(\rho)]^2 \rho \, d\rho}{8\pi [\int_0^1 g(\rho) \, q(\rho) \, d\rho]^2}$$

where

(14) $$q(\rho) = \sum_{k=1}^\infty k^2 |a_k|^2 \rho^{2k-1} .$$

We define

(15) $$Q(\rho) = (1/2) \sum_{k=1}^\infty k |a_k|^2 \rho^{2k}$$

so that

(16) $$Q'(\rho) = q(\rho), \qquad Q(0) = 0 .$$

Using this and (6), we obtain

(17) $$\int_0^1 g(\rho) \, q(\rho) \, d\rho = - \int_0^1 g'(\rho) \, Q(\rho) \, d\rho .$$

Now

$$(18) \qquad \left[\int_0^1 g'(\rho)\, Q(\rho)\, d\rho\right]^2 \leq \int_0^1 [g'(\rho)]^2 \rho\, d\rho \int_0^1 [Q(\rho)]^2 \rho^{-1}\, d\rho .$$

By (17) and (18), the minimum of the right hand side of (13) is attained when

$$(19) \qquad g'(\rho)\, \rho^{1/2} = Q(\rho)\, \rho^{-1/2} .$$

This choice of $g(\rho)$ changes (13) into

$$(20) \qquad P \geq 8\pi \int_0^1 [Q(\rho)]^2 \rho^{-1}\, d\rho$$

which, by (15) yields

$$(21) \qquad P \geq \pi \sum_{k=1}^{\infty} \sum_{l=1}^{\infty} \frac{kl}{k+l} |a_k|^2 |a_l|^2 .$$

We see easily that (21) gives a better lower bound for P than (11) by comparing the coefficient of $|a_k|^2 |a_l|^2$ in both expansions.]

5.8. Conformal mapping. Estimate for Λ. We consider the same conformal mapping as in the foregoing section and assume again 5.7 (5). Then

$$(1) \qquad \iint f^2\, dx\, dy = 2\pi \int_0^1 [g(\rho)]^2 \sum_{k=1}^{\infty} k^2 |a_k|^2 \rho^{2k-1}\, d\rho ;$$

see 5.7 (7). Combining (1), 5.1 (3) and 5.7 (9), we obtain

$$(2) \qquad \Lambda^2 \leq \frac{\int_0^1 [g'(\rho)]^2 \rho\, d\rho}{\int_0^1 [g(\rho)]^2 \sum_1^{\infty} k^2 |a_k|^2 \rho^{2k-1}\, d\rho}$$

We choose $g(\rho)$ according to 5.6 (3); this choice satisfies 5.7 (6). We obtain so

$$(3) \qquad \Lambda^2 \leq \frac{j^2}{|a_1|^2 + \lambda_2 |a_2|^2 + \lambda_3 |a_3|^2 + \dots} ;$$

we have introduced the abbreviation

$$(4) \qquad \lambda_k = k^2 \int_0^1 [J_0(j\rho)]^2 \rho^{2k-1} d\rho \Big/ \int_0^1 [J_0(j\rho)]^2 \rho \, d\rho \; .$$

We observe that $\lambda_k > 0$ and derive from (3) the weaker but more concise inequality

$$(5) \qquad \Lambda < j/\dot{r}$$

valid unless C is a circle.

5.9. Examples and comments. In the previous sections we have obtained various lower bounds for P and upper bounds for Λ. Each of these bounds depends on the choice of a suitable point within the curve C. In sect. 5.5 and 5.6 we have to choose a point a from which the perpendiculars to a variable tangent are drawn. In sect. 5.7 and 5.8 we have to choose the point a, the image of the center of the circle. In the estimates 5.5 (9), 5.6 (4), 5.7 (12) and 5.8 (5) we assume implicitly that the choice has been made, namely so as to give B_a its minimum value B and r_a its maximum value $\dot{r}$. This choice is actually not easy to make, yet these estimates remain true a fortiori if B is replaced by B_a or $\dot{r}$ by r_a. If D is convex and has a center of symmetry, the most advantageous position for a seems to be just that center. [That this is actually so, has been proved for B_a by Aissen, see 1, and for r_a by Haegi, see 2 and sect. 7.5 (b).] At any rate, this will be our choice in the following examples.

(a) For an ellipse with semi-axes a and b we easily obtain, taking the center as the origin for h,

$$(1) \qquad B_0 = \int h^{-1} ds = \pi(a^2 + b^2)/(ab) \quad .$$

On substituting (1), both sides of 5.5 (9) turn out to be equal. In fact, in deriving 5.5 (9), we aimed at a lower bound that should coincide with the true value for an arbitrary ellipse.

On substituting (1), 5.6 (4) gives an upper bound for the principal frequency Λ of an ellipse. We present it here along with the lower bound given by the theorem of Rayleigh-Faber-Krahn (1.21, entry 15):

$$(2) \qquad \frac{j^2}{ab} \leqq \Lambda^2 \leqq \frac{j^2}{2}\left(\frac{1}{a^2} + \frac{1}{b^2}\right) \; .$$

The two bounds, the arithmetic and the geometric means of $(j/a)^2$ and $(j/b)^2$, coincide, of course, with each other and with the correct value in the case of the circle when a=b. In the case of a very flat ellipse, when b/a is very small, the lower bound for Λ is almost 100% too low, but the upper bound is only about 8% too high (in the ratio $j\sqrt{2} : \pi$).

(b) For a rectangle with sides a and b we obtain, taking the center as the origin for h,

(3) $$B_0 = \int h^{-1}\, ds = 4(a^2 + b^2)/(ab).$$

On substituting (3), 5.5 (9) and 5.6 (4) yield

(4) $$P > \frac{a^3 b^3}{4(a^2 + b^2)},$$

(5) $$\Lambda^2 < 2j^2 \left(\frac{1}{a^2} + \frac{1}{b^2}\right),$$

respectively. It is remarkable that the bound for Λ given by (5) has a fixed proportion to the true value of Λ ($j\sqrt{2} : \pi$ or roughly 108:100), which is independent of the shape of the rectangle.

The bounds 5.7 (12) and 5.8 (5), derived by conformal mapping, are not too explicit, but we can obtain easily numerical values in the two extreme cases: for the square, when a = b, and for a very flat rectangle, when b/a is very small. The following short table compares the approximation by conformal mapping and by similar level lines (that is, by (4), (5)); it gives approximate values of the percentage errors:

	Square		Flat rectangle	
	P	Λ	P	Λ
Similar level lines:	-11%	+ 8%	- 25%	+ 8%
Conformal mapping:	- 5%	+0.36%	-100%	+20%

Closer bounds are obtained by conformal mapping in the case of the square, and by similar level lines in the case of the flat rectangle. The bound for Λ obtained by conformal mapping shows a remarkably good approximation in the case of the square.

(c) Let $\underline{D}$ be convex and ρ the radius of the inscribed circle, that is, of the largest circle contained in $\underline{D}$. Choosing the center of this circle as the origin for h, we obtain

(6) $$B \leq \int \rho^{-2} h\, ds = 2A/\rho^2$$

where A represents, as usual, the area of the domain $\underline{D}$. Equality is attained in (6) if $\underline{D}$ is a polygon circumscribed about a circle.

By using (6), we conclude from 5.5 (9) that

$$P \geqq \frac{1}{2} A \rho^2 . \tag{7}$$

Equality is attained in (7) when $\underline{D}$ is a circle. By using (6), we obtain only the following trivial result from 5.6 (4): The principal frequency of $\underline{D}$ cannot be higher than that of the largest circle contained in $\underline{D}$.

[5.9 A. Prescribed level lines. Estimate of P. (a) A more general treatment of our problem is possible by prescribing a certain set of curves as the level curves of the function f in 5.1 (2). Let $\underline{C}_\rho$ denote an arbitrary set of curves, $0 \leqq \rho \leqq 1$, so that $\underline{C}_0$ is an interior point (or a finite set of interior points) of the domain $\underline{D}$ bounded by the curve $\underline{C}$, and $\underline{C}_1 = \underline{C}$. We assume that $\underline{C}_\rho$ is either a simple curve or a finite system of mutually exclusive simple curves. Finally we assume that the curves constituting $\underline{C}_\rho$ are in the interior of $\underline{C}_{\rho'}$, if $\rho < \rho'$, and the $\underline{C}_\rho$ fill $\underline{D}$ completely.

We introduce the integrals

$$\int_{\underline{C}_\rho} |\text{grad } \rho| \, ds = \lambda(\rho) , \qquad \int_{\underline{C}_\rho} |\text{grad } \rho|^{-1} \, ds = \mu(\rho) \tag{1}$$

where ds is the arc element of $\underline{C}_\rho$. The meaning of $\mu(\rho)$ is clear: we have $\mu(\rho) = A'(\rho)$ where $A(\rho)$ is the total area bounded by $\underline{C}_\rho$. Indeed, the area element of the infinitesimal ring-shaped domain between $\underline{C}_\rho$ and $\underline{C}_{\rho+d\rho}$, $d\rho > 0$, can be written in two different ways:

$$A(\rho + d\rho) - A(\rho) = A'(\rho)\, d\rho \tag{2}$$

and

$$\int_{\underline{C}_\rho} ds \, . \, dn \tag{3}$$

where dn is the piece of the normal of $\underline{C}_\rho$ between $\underline{C}_\rho$ and $\underline{C}_{\rho+d\rho}$. But $|\text{grad } \rho| = d\rho/dn$ so that

$$A'(\rho) = \int_{\underline{C}_\rho} ds \, . \, |\text{grad } \rho|^{-1} = \mu(\rho) . \tag{4}$$

Now we choose in 5.1 (2):

$$f = g(\rho) , \qquad g(1) = 0 . \tag{5}$$

The contribution of the domain between $\underline{C}_\rho$ and $\underline{C}_{\rho+d\rho}$ to Dirichlet's integral is:

$$\int_{\underline{C}_\rho} |\text{grad } f|^2 \; ds \, . \, dn \tag{6}$$

where dn has the previous meaning. In view of $dn = |\text{grad } \rho|^{-1} \, d\rho$ we can write (6) as follows:

$$[g'(\rho)]^2 \int_{\underline{C}_\rho} |\text{grad } \rho|^2 \, . \, |\text{grad } \rho|^{-1} \, ds \, d\rho = [g'(\rho)]^2 \, \lambda(\rho) \, d\rho \, . \tag{7}$$

On the other hand, the contribution of the same infinitesimal domain to the integral in the denominator of 5.1 (2) will be:

$$g(\rho) \int_{\underline{C}_\rho} ds \, . \, dn = g(\rho) \int_{\underline{C}_\rho} |\text{grad } \rho|^{-1} \, ds \, d\rho = g(\rho) \, \mu(\rho) \, d\rho \tag{8}$$

so that we have, according to 5.1 (2),

$$\frac{1}{P} \leqq \frac{\int_0^1 [g'(\rho)]^2 \, \lambda(\rho) \, d\rho}{4[\int_0^1 g(\rho) \, \mu(\rho) \, d\rho \,]^2} \, . \tag{9}$$

Introducing the function $M(\rho)$ by the conditions

$$M'(\rho) = \mu(\rho) \, , \qquad M(o) = 0 \, , \tag{10}$$

we have

$$\int_0^1 g(\rho) \, \mu(\rho) \, d\rho = - \int_0^1 g'(\rho) \, M(\rho) \, d\rho \, . \tag{11}$$

By Schwarz's inequality

$$[\int_0^1 g'(\rho) \, M(\rho) \, d\rho \,]^2 \leqq \int_0^1 [g'(\rho)]^2 \, \lambda(\rho) \, d\rho \int_0^1 [M(\rho)]^2 [\lambda(\rho)]^{-1} \, d\rho \tag{12}$$

and equality holds when

$$g'(\rho) \, [\lambda(\rho)]^{1/2} = M(\rho) \, [\lambda(\rho)]^{-1/2} \, . \tag{13}$$

This yields for the torsional rigidity the following inequality:

$$P \geqq 4 \int_0^1 [M(\rho)]^2 \, [\lambda(\rho)]^{-1} \, d\rho \; . \tag{14}$$

The meaning of $M(\rho)$ is clear:

$$M(\rho) = A(\rho) \; . \tag{15}$$

(b) In the special case that we considered in 5.7 (c), the curve $\underline{C}_\rho$ was the image of the circle $|\zeta| = \rho$ in the conformal mapping 5.7 (1). The area of the domain bounded by $\underline{C}_\rho$ is

$$A(\rho) = \pi \sum_{k=1}^{\infty} k|a_k|^2 \rho^{2k} = M(\rho)$$

so that

$$M(\rho) = 2\pi Q(\rho) \; , \tag{16}$$

see 5.7 (15). Moreover we have on $\underline{C}_\rho$:

$$|\text{grad}\, \rho| = \frac{d\rho}{dn} = \frac{d\rho}{|z'(\zeta)|\, d\rho} = \frac{1}{|z'(\zeta)|} \; , \tag{17}$$

$$ds = |dz| = |z'(\zeta)|\, |d\zeta|$$

so that

$$\lambda(\rho) = \int_{\underline{C}_\rho} |\text{grad}\, \rho| \, ds = \int |d\zeta| = 2\pi\rho \; . \tag{18}$$

In view of (16) and (18), (14) yields the inequality 5.7 (20).

(c) In order to prepare a new application of (14) we consider a multiply connected domain $\underline{D}$. The boundary of $\underline{D}$ consists of a finite number of simple curves, one of them (the outer boundary) contains all the others (the inner boundaries); the inner boundaries are mutually exclusive curves. We introduce the following notation. We call Δ the simply connected domain the only boundary curve of which is the outer boundary curve of $\underline{D}$. Therefore, Δ contains $\underline{D}$ and also the "holes" of $\underline{D}$, that is, the simply connected domains each of which is bounded by one of the inner boundary curves of $\underline{D}$.

The variational problem discussed in sect. 5.2 that leads to the definition of P by the inequality 5.1 (2) can be extended to multiply connected domains in two ways, both of which admit a physical interpretation. In both cases we consider a function f which is sufficiently smooth in Δ, but arbitrary otherwise except for certain conditions which we shall specify immediately.

First, we assume that $f > 0$ in the interior of $\underline{D}$, but $f = 0$ along the boundary of Δ and in each hole (and, therefore, along the whole boundary of $\underline{D}$). Admitting such functions f, we seek the minimum $1/P'$ of the left hand side of 5.1 (2); the integrals are extended over Δ or, which comes to the same, over $\underline{D}$. The existence of P' is proved substantially by the argument of sect. 5.2.

Second, we assume that $f > 0$ in the interior of $\underline{D}$, $f = 0$ along the boundary of Δ (the outer boundary curve of $\underline{D}$) and that f remains constant in each of the holes (and, therefore, along each inner boundary curve of $\underline{D}$); the constants belonging to different holes may be different, however. Admitting such functions f, we seek the minimum $1/P''$ of the left hand side of 5.1 (2); the integrals are extended over Δ (not over $\underline{D}$, which is important for the denominator). In order to prove the existence of P'', the argument of sect. 5.2 needs essential additions; see 25.

Obviously, as there is more freedom of choice in the second problem, $1/P'' \leqq 1/P'$. In fact, it can be shown that

$$P'' > P' . \tag{19}$$

For example, if $\underline{D}$ is the annulus between two concentric circles with radii a and b respectively, $a < b$,

$$P' = \frac{\pi}{2}\left[b^4 - a^4 - \frac{(b^2 - a^2)^2}{\log(b/a)}\right] ,$$

$$P'' = \frac{\pi}{2}\,[b^4 - a^4]$$

and the inequality (19) is obviously verified.

P'' is proportional to the torsional rigidity of a beam with cavities the cross-section of which is $\underline{D}$, while P' is proportional to the discharge of a viscous fluid flowing through a compound pipe the cross-section of which is $\underline{D}$. For a simply connected domain the quantities P' and P'' coincide with P.

(d) We derive now a lower bound for the discharge P'.

By the definition explained under (c), the inequality 5.1 (2) holds if the integrals are extended over the multiply connected domain $\underline{D}$, $f = 0$ along the entire boundary of $\underline{D}$, and P' is substituted for P. We obtain hence (14) with P' instead of P.

Let a be an arbitrary interior point of $\underline{D}$. We denote by G the Green function of the domain $\underline{D}$ with respect to the point a. This function is regular harmonic in $\underline{D}$ except at a; in the neighborhood of a we have

(20) $$G = \log \frac{1}{r} - h$$

where h is regular harmonic. Moreover $G = 0$ holds on all the boundary curves of $\underline{D}$. The value of h at the point a is $\log (1/r_a)$ where r_a is called the inner radius of the multiply connected domain $\underline{D}$ with respect to a (32, p. 335). This concept of the inner radius is more general than that given in sect. 1.3 where only simply connected domains were considered.

The Green function G varies between 0 and $+\infty$ in the domain $\underline{D}$. We choose for $\underline{C}_\rho$ the curve (or system of curves)

(21) $$G = \log \frac{1}{\rho} .$$

The area $A(\rho)$ of the domain bounded by $\underline{C}_\rho$ satisfies the inequality (29)

(22) $$A(\rho) \geq \pi \rho^2 r_a^2$$

so that

(23) $$M(\rho) = A(\rho) \geq \pi \rho^2 r_a^2 .$$

On the other hand

(24) $$\lambda(\rho) = \int_{\underline{C}_\rho} \frac{d\rho}{dn}\, ds = -\rho \int_{\underline{C}_\rho} \frac{\partial G}{\partial n}\, ds = 2\pi\rho$$

so that from (14)

(25) $$P' \geq 4 \int_0^1 (\pi \rho^2 r_a^2)^2 (2\pi\rho)^{-1}\, d\rho = \frac{\pi}{2} r_a^4$$

follows. Thus 5.7 (12) holds for the multiply connected domain $\underline{D}$ with P' or P" substituted for $\underline{D}$, see (19).

(e) Following the argument under (a) we derive

(26) $$\Lambda \leq \frac{\int_0^1 [g'(\rho)]^2 \lambda(\rho)\, d\rho}{\int_0^1 [g(\rho)]^2 \mu(\rho)\, d\rho}$$

from 5.1 (3) as we have derived (9) from 5.1 (2). In (26), as in (9), we have to choose a function $g(\rho)$ with $g(1) = 0$ that minimizes the right hand side. The Euler-Lagrange equations corresponding to (9) and (26) are

$$[\lambda(\rho)\, g'(\rho)]' = c\,\mu(\rho) ,$$

$$[\lambda(\rho)\, g'(\rho)]' = c\,\mu(\rho)\, g(\rho) ,$$

respectively, where c is a suitable constant. The first can be solved explicitly, see (13), but not the second, and this is the reason that we cannot follow up the treatment of Λ by the method of the prescribed level lines to a stage that would correspond to (14).]

[5.9 B. A direct relation between torsional rigidity and the vibrations of a membrane. The complete solution of the problem of a vibrating membrane consists in finding an infinite system of eigenfunctions

(1) $$\varphi_1, \quad \varphi_2, \quad \varphi_3, \ldots \quad \varphi_n, \ldots$$

and corresponding eigenvalues

(2) $$\lambda_1, \quad \lambda_2, \quad \lambda_3, \ldots \quad \lambda_n, \ldots$$

which are connected by the relations

(3) $$\nabla^2 \varphi_n + \lambda_n^2 \varphi_n = 0 \qquad \text{in } D ,$$

(4) $$\varphi_n = 0 \qquad \text{on } C,$$

(5) $$\iint \varphi_k\, \varphi_l\, dx\, dy = 0 \qquad \text{if } k \neq l ,$$

(6) $$0 < \lambda_1 \leqq \lambda_2 \leqq \lambda_3 \leqq \ldots .$$

It is understood that φ_n does not vanish identically. It is assumed that the system (1) of orthogonal functions is complete in the simply connected domain D. In the notation of sect. 5.3

(7) $$\lambda_1 = \Lambda , \qquad \varphi_1 = w$$

and this is an important point of contact with the foregoing investigation. Our present aim is to express P, the torsional rigidity of D, in terms of the sequences (1) and (2).

(a) We assume that the function v considered in sect. 5.2 has an expansion of the form

(8) $$v = \sum a_n \varphi_n .$$

The a_n are constants (Fourier-coefficients), the summation Σ is extended over $n = 1, 2, 3, \ldots$. In view of (4), the form of the expansion (8) guarantees the fulfillment of the boundary condition 5.2 (2). It remains to satisfy the differential equation 5.2 (1). Substituting for v the expansion (8), we obtain in view of (3)

(9) $$\nabla^2 v = \sum a_n \nabla^2 \varphi_n = - \sum \lambda_n^2 a_n \varphi_n = -2 .$$

We assume the expansion

(10) $$1 = \sum c_n \varphi_n .$$

By comparing (9) and (10) and using the orthogonality relations (5), we find

(11) $$\lambda_n^2 a_n = 2 c_n = 2 \iint \varphi_n \, dx\, dy \Big/ \iint \varphi_n^2 \, dx\, dy .$$

This determines the expansion (8) of the stress function v. Now (8), (11) and 5.2 (4) give

(12) $$P = 4 \sum \frac{[\iint \varphi_n \, dx\, dy]^2}{\lambda_n^2 \iint \varphi_n^2 \, dx\, dy} .$$

(b) We may be led to (12) by the variational definition 5.1 (2) of P. In fact, following the Rayleigh-Ritz method, we consider a finite linear combination of the functions (1) with indeterminate coefficients u_n:

(13) $$f = \sum_{n=1}^{N} u_n \varphi_n .$$

In view of (4), this f satisfies the boundary condition 5.1 (1). We wish to minimize the left hand side of 5.1 (2) by a suitable choice of the u_n. Now

(14) $$\iint \operatorname{grad} \varphi_k \cdot \operatorname{grad} \varphi_l \, dx\, dy$$

$$= - \iint \varphi_k \nabla^2 \varphi_l \, dx\, dy - \int \varphi_k \frac{\partial \varphi_l}{\partial n_i} \, ds$$

$$= \lambda_l^2 \iint \varphi_k \varphi_l \, dx\, dy$$

$$= \begin{cases} \lambda_l^2 \iint \varphi_l^2 \, dx\, dy & \text{if } k = l , \\ 0 & \text{if } k \neq l . \end{cases}$$

The line integral is along $\underline{C}$; we used (3), (4) and (5). Therefore, see (13),

$$\iint |\text{grad } f|^2 \, dx\, dy = \sum_{n=1}^{N} u_n^2 \cdot \lambda_n^2 \iint \varphi_n^2 \, dx\, dy$$

and 5.1 (2) becomes

$$(15) \qquad P \geq \frac{4\left[\sum_{n=1}^{N} u_n \iint \varphi_n \, dx\, dy\right]^2}{\sum_{n=1}^{N} u_n^2 \lambda_n^2 \iint \varphi_n^2 \, dx\, dy} .$$

Cauchy's inequality shows that the best choice that maximizes the right hand side of (15) is

$$u_n = \frac{\iint \varphi_n \, dx\, dy}{\lambda_n^2 \iint \varphi_n^2 \, dx\, dy}$$

and this yields

$$(16) \qquad P \geq 4 \sum_{n=1}^{N} \frac{\left[\iint \varphi_n \, dx\, dy\right]^2}{\lambda_n^2 \iint \varphi_n^2 \, dx\, dy} .$$

We have obtained an essential part of (12).

(c) From (10) and (11) we obtain an expression for the area of $\underline{D}$:

$$(17) \qquad A = \sum_{n=1}^{\infty} \frac{\left[\iint \varphi_n \, dx\, dy\right]^2}{\iint \varphi_n^2 \, dx\, dy} .$$

We can compare (12) with (17). Combining both with (6) and (7) we obtain the inequality 5.4 (2).

In (a) we derived a complete solution of the problem of torsion from a complete solution of the problem of the vibrating membrane, obtaining v and P. It is instructive to carry through the computation of P by (12) in the best known cases.

(d) We consider the rectangle with the four corners (o,o), (a,o), (a,b), (o,b). The eigenfunctions and eigenvalues are of the form

$$\varphi = \sin \frac{\pi k x}{a} \sin \frac{\pi l y}{b} ,$$

$$\lambda^2 = \pi^2 \left(\frac{k^2}{a^2} + \frac{l^2}{b^2} \right)$$

where k and l are positive integers. We find

$$\iint \varphi \, dx \, dy = \frac{4 \, ab}{\pi^2 \, kl}$$

if both k and l are odd numbers; in all other cases the integral vanishes. In all cases

$$\iint \varphi^2 \, dx \, dy = ab/4 \quad .$$

Substituting these values in (12), we obtain for the torsional rigidity of the rectangle with sides a and b

$$P = \frac{256}{\pi^6} \, a \, b \sum \sum \frac{1}{k^2 l^2 \left(\frac{k^2}{a^2} + \frac{l^2}{b^2} \right)} \quad ; \tag{18}$$

the double summation is extended over all <u>odd</u> positive values, 1, 3, 5, ... of k and l. The leading term k = l = 1 gives by itself a better lower estimate than 5.9 (4). Using classical formulas (the expansion of tan z in partial fractions), we can easily transform the expression (18) into the well known expression for the torsional rigidity of a rectangle given by Saint-Venant.

(e) We consider the isosceles right triangle with corners (o,o), (π,o), (π, π). The eigenfunctions and eigenvalues are of the form

$$\varphi = \sin kx \, \sin ly - \sin lx \, \sin ky \, ,$$

$$\lambda^2 = k^2 + l^2$$

where k and l are different positive integers. We find that $\iint \varphi \, dx \, dy$ vanishes unless one of the two numbers k and l is even and the other odd. We may assume k = 2m, l = 2n - 1 and find from (12) for the torsional rigidity of the triangle that is one half of a square with side a

$$P = \frac{1024}{\pi^6} a^4 \sum_{m=1}^{\infty} \sum_{n=1}^{\infty} \frac{m^2}{(2n-1)^2 \, [4m^2 - (2n-1)^2][16m^4 - (2n-1)^4]} \quad ; \tag{19}$$

we pass from the special case a = π to a general a by considering the dimension of P. Using classical formulas, we can effect the summation with respect to m and transform (19) into the better known series

$$P = a^4 \left[\frac{1}{12} - \frac{16}{\pi^5} \sum_{n=1}^{\infty} \frac{1}{(2n-1)^5} \coth \frac{(2n-1)\pi}{2} \right] \tag{20}$$

(f) We consider the unit circle. It is easily seen that $\iint \varphi \, dx\, dy$ vanishes if φ has a radial nodal line. We easily obtain from (12) that the torsional rigidity of the unit circle is

$$P = 16\pi \sum_{n=1}^{\infty} j_n^{-4} \tag{21}$$

where $j_1, j_2, j_3, \ldots$ are the positive roots of the Bessel function $J_0(x)$; especially $j_1 = j$, see 1.3 (1). Expanding the logarithmic derivative of $J_0(x)$ in powers of x, we find that (21) reduces to $P = \pi/2$.]

Applications of the Inclusion Lemma

5.10. The inclusion lemma. We consider a convex domain D which we assume to be bounded and closed. Then there are two points I and K in D the distance of which is a maximum. (See Fig. 3; the length of the segment IK is called the diameter of D.) It follows easily, that the perpendiculars to IK through the points I and K (AB and DC in Fig. 3) are supporting lines of D. Drawing the two supporting lines parallel to IK (through J and L in Fig. 3) we obtain a rectangle ABCD which contains D. On the other hand D, being convex, contains the quadrilateral IJKL. Since IK is parallel to AD and BC, the area of the quadrilateral IJKL is one half of the area of the rectangle ABCD. We have obtained so our first result:*

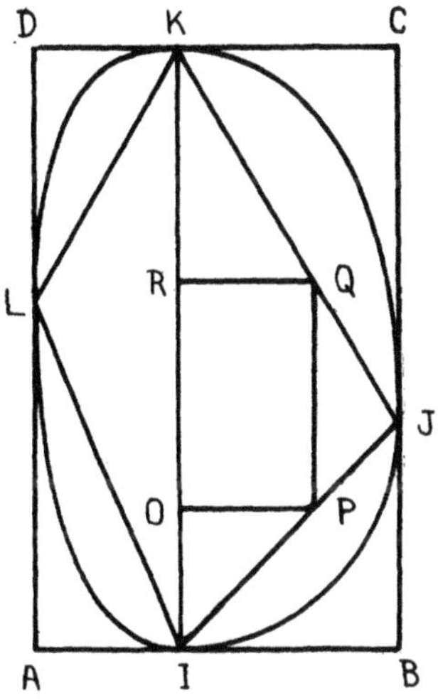

Fig. 3

LEMMA I. About any convex domain D with area A we can circumscribe a rectangle with area A_2 so that

$$A_2 \leqq 2A . \tag{1}$$

It can be shown, by the example of the triangle, that 2 is the best possible (the lowest) constant with which Lemma I holds but we do not enter into details. We keep on considering the situation represented by Fig. 3. The point I divides the base of the rectangle ABCD into two segments. One of these, say IB, is longer (not shorter) than the other:

$$IB \geqq AB/2 . \tag{2}$$

* This derivation is different from our original one, and is due to Professor H. Rademacher.

In the triangle IJK we draw a parallel PQ to the base IK such that

$$(3) \qquad PQ = IK/3 = BC/3 .$$

It follows from this and from (2) that

$$(4) \qquad OP = 2IB/3 \geqq AB/3 .$$

Observe that the rectangle OPQR is wholly contained in the given convex domain $\underline{D}$ and that, by (3) and (4), no side of OPQR is less than one third of the corresponding side of ABCD. We obtain so:

LEMMA II. Being given an arbitrary convex domain $\underline{D}$, we can find two rectangles R_1 and R_2 with the following properties:
R_1 is contained in $\underline{D}$,
R_2 contains $\underline{D}$,
R_1 and R_2 are similar to each other and similarly situated and, if we let $[R_2]/[R_1]$ denote the ratio of corresponding lines in R_2 and R_1,

$$[R_2]/[R_1] \leqq 3 .$$

Is 3 the best possible bound, or could we improve Lemma II by substituting for 3 a smaller number? We do not discuss this question but pass to the next result.

LEMMA III. Being given an arbitrary convex domain $\underline{D}$, we can find two ellipses E_1 and E_2 with the following properties:
E_1 is contained in $\underline{D}$,
E_2 contains $\underline{D}$,
E_1 and E_2 are similar to each other and similarly situated and, if we let $[E_2]/[E_1]$ denote the ratio of corresponding lines in E_2 and E_1,

$$[E_2] / [E_1] \leqq q$$

where q is an absolute constant independent of $\underline{D}$.

In order to prove Lemma III we start from Lemma II. Performing an affine transformation, if necessary, we assume that R_1 and R_2 are squares. Inscribing a circle in R_1 and circumscribing another circle about R_2, we obtain Lemma III with value $q = 3\sqrt{2}$. (This value is much too high but we do not need here the best value of q.)*

If the domain $\underline{D}$ has a center of symmetry, a line of maximum distance

* See, F. Behrend, Über die kleinste umgeschriebene und die grösste eingeschriebene Ellipse eines konvexen Bereiches, Math. Annalen, vol. 115 (1938) pp. 379-411. -F. John, Extremum problems with inequalities as subsidiary conditions, Studies and Essays presented to R. Courant, 1948, pp. 187-204.

(as IK in Fig. 3) must pass through this center. This remark leads to an appropriate modification of Fig. 3, which eventually yields the following:

> LEMMA IV. If the domain D has a center of symmetry, the rectangles R_1 and R_2 and ellipses E_1 and E_2, mentioned in Lemmas II and III, can be so chosen that R_1, R_2, E_1, E_2 and D have the same center of symmetry.

Using the term "inclusion lemma," we shall refer jointly to Lemmas I, II, III and IV, the contents of which we could express rather vaguely but suggestively by saying that an arbitrary convex curve is only "boundedly different" from a suitable rectangle or a suitable ellipse.

5.11. Applications. It will be enough to consider in detail just one example.

(a) De Saint-Venant's approximate formula for torsional rigidity. We wish to estimate the quantity

$$PIA^{-4}$$

for an arbitrary convex domain D, knowing that its value is

$$(2\pi)^{-2}$$

for an arbitrary ellipse. Being given D, we consider the ellipses E_1 and E_2 mentioned in Lemma III. Let A_1, I_1, P_1, A_2, I_2 and P_2 denote the quantities which are connected in the same way with E_1 and E_2, respectively, as A, I and P are with D. Then

$$I_1 \leqq I \leqq I_2, \qquad P_1 \leqq P \leqq P_2,$$

$$A_1 \leqq A \leqq A_2 \leqq q^2 A_1$$

and therefore

$$PIA^{-4} \leqq P_2 I_2 A_1^{-4} \leqq P_2 I_2 q^8 A_2^{-4} = q^8 (2\pi)^{-2} ,$$

$$PIA^{-4} \geqq P_1 I_1 A_2^{-4} \geqq P_1 I_1 q^{-8} A_1^{-4} = q^{-8} (2\pi)^{-2} .$$

We express the essential content of the two last inequalities by saying that PIA^{-4} is contained between positive bounds for an arbitrary convex domain. This result was announced in 1.17 and listed under entry 19 in 1.21.

Good numerical approximations to the best bounds would be of some practical importance, but they are probably not easy to obtain. The numerical values yielded by the foregoing argument are probably far from the best possible values.

(b) *Further examples*. We summarize the foregoing argument: *From the fact that* PIA^{-4} *has a finite upper* (or lower) *bound for an arbitrary ellipse, we infer the same for an arbitrary convex domain by using the inclusion lemma*.

The same line of argument leads to many similar results several of which are listed in the table of 1.21. Sometimes we start from ellipses (entries 12, 18, 19), sometimes from rectangles (entries 13, 14, 20, 21). To establish the boundedness of the considered quantity for the chosen particular class of domains (ellipses or rectangles) is often more difficult than it was under (a).*

We do not enter into details here, but we mention a case not listed in 1.21. We consider *D* as covered with matter of uniform surface density 1. Let I' and I" denote the moments of inertia about those two principal axes passing through the center of gravity which lie in the plane of *D*. (Therefore, with our notation explained in 1.3, I' + I" = I.) Then (E. Nicolai; see *4*)

$$P \leqq \frac{4I'I''}{I' + I''} \tag{1}$$

and there is equality for an arbitrary ellipse. For convex domains the ratio of the right hand side of (1) to the left hand side is *bounded* also *from above*; this we infer from Lemma III. Yet for non-convex domains the ratio is unbounded as examples show.

(c) [Let ρ and R denote the radius of the inscribed and circumscribed circle of the convex domain *D*, respectively. That is, ρ is the maximum of the radii of the circles contained in the closed domain *D* and R the minimum of the radii of the circles containing *D*. For instance, if *D* is an ellipse, its area $A = \pi \rho R$. On the basis of the inclusion lemma, it is easy to show that each functional in the first row of the following table is "comparable" with the corresponding functional in the second row:

L	$\bar{r}$	C	$\bar{I}$	$\bar{A}$	$\bar{P}$	$\bar{\Lambda}$	$\dot{r}$
R	R	$R[\log(1 + \frac{R}{\rho})]^{-1}$	$R^{3/4}\rho^{1/4}$	$R^{1/2}\rho^{1/2}$	$R^{1/4}\rho^{3/4}$	ρ	ρ

Also B is comparable with $R\rho^{-1}$. Two functionals of a convex plane domain *D* are called *comparable* if their quotient remains between fixed positive bounds independent of the shape and size of *D*. See M. Aissen, *1*.

* The discussion of entry 13 can be based on H. Davenport and G. Pólya, On the product of two power series, Canadian Journal of Math., vol. 1 (1949) pp. 1-5.

The functionals listed in the table have, unlike B, the dimension of a line. The second row of the table shows the reason for arranging them in the displayed order. It is worth noticing that this order is the same as the order of the corresponding variational formulas in the table of sect. 1.33.

As an application we mention that $\dot{\bar{r}}\bar{r}A^{-1}$ remains bounded for all convex domains; compare 1.23. Good bounds (although not the best bounds) have been established by Haegi, see 9.

Of course, many approximate formulas can be made up which behave similarly as Saint-Venant's 1.17 (1) discussed under (a). The following has been proposed by Aissen (1):

$$(2) \qquad P \sim \frac{A^2}{\pi} \frac{R\rho}{R^2 + \rho^2}$$

and is exact for an arbitrary ellipse. We mention the analogous

$$(3) \qquad \Lambda \sim \frac{2\pi R}{A}$$

which is exact for an arbitrary rectangle.]

Applications of Conformal Mapping

5.12. An assumption. We consider again the conformal mapping 5.7 (1), substituting a_0 for a :

$$(1) \qquad z = a_0 + a_1\zeta + a_2\zeta^2 + \ldots + a_n\zeta^n + \ldots \quad .$$

The series (1) converges for $|\zeta| < 1$. For the sake of convenience, however, we shall assume more, namely that the derived series converges absolutely for $|\zeta| \leq 1$ or, what is the same, that the numerical series

$$(2) \qquad |a_1| + 2|a_2| + \ldots + n|a_n| + \ldots$$

converges. This assumption underlies our work till section 5.17 inclusively. In section 5.18, however, we shall show that our main result remains valid under a less restrictive and more natural condition.

5.13. The boundary value problem of torsion. This problem consists in solving the differential equation 5.2 (1') for Φ under the boundary condition 5.2 (2'). By 5.2 (1'), the function Φ should be harmonic in the interior of D, and, therefore, it should be the real part of a certain analytic function

(1) $$F = \Phi + i\Psi$$

of the complex variable $z = x + iy$ which is regular in the interior of D. The function F is determined by Φ except for a purely imaginary additive constant. As z itself is an analytic function of ζ, represented by 5.12 (1), F, depending on z, is also an analytic function of ζ in the circle $|\zeta| < 1$ and is represented there by a power series

(2) $$F = u_0 + u_1\zeta + u_2\zeta^2 + \ldots + u_n\zeta^n + \ldots$$

We dispose of the arbitrary additive constant contained in F by taking u_0 real, or

(3) $$u_0 = \bar{u}_0 .$$

Finally, our problem of determining Φ turns out to be equivalent to determining the sequence $u_0, u_1, u_2, \ldots$.

We attempt to satisfy the condition 5.2 (2') in supposing that $\sum u_n$ is absolutely convergent. Thus, we should have, by 5.2 (2'), 5.12 (1), (1), (2) and (3), for $|\zeta| = 1$,

(4) $$2\Phi = \sum_{n=0}^{\infty} (u_n\zeta^n + \bar{u}_n\bar{\zeta}^n)$$

$$= 2u_0 + \sum_{n=1}^{\infty} (u_n\zeta^n + \bar{u}_n\bar{\zeta}^n)$$

$$= |z|^2$$

$$= \sum_{k=0}^{\infty} a_k\zeta^k \sum_{l=0}^{\infty} \bar{a}_l\zeta^{-l}$$

$$= \sum_{n=-\infty}^{\infty} b_n\zeta^n .$$

We have introduced the abbreviation

(5) $$b_n = \sum_{k-l=n} a_k\bar{a}_l = \sum_{l=0}^{\infty} a_{l+n}\bar{a}_l$$

and put

(6) $$a_m = 0 \qquad \text{if } m < 0 .$$

It follows from (5) that

(7) $$\bar{b}_n = \sum_{k-l=n} \bar{a}_k a_l = \sum_{l-k=n} \bar{a}_l a_k = b_{-n} .$$

We satisfy the condition (4) by choosing

(8) $$2u_0 = b_0, \quad u_n = b_n, \quad \bar{u}_n = b_{-n} \quad (n = 1,2,3,\ldots) .$$

(This is the only admissible choice, by the uniqueness theorem of Fourier series, only the simplest case of which is needed here.)

Using the values (8) just obtained and putting

(9) $$\zeta = \rho e^{i\varphi}$$

with real ρ and φ, $\rho \geq 0$, we finally obtain, in view of (4), (5) and (7),

(10) $$2\Phi = b_0 + \sum_{n=1}^{\infty} (b_n \zeta^n + \bar{b}_n \bar{\zeta}^n)$$

$$= \sum_{-\infty}^{\infty} b_n \rho^{|n|} e^{in\varphi}$$

$$= \sum_{k=0}^{\infty} \sum_{l=0}^{\infty} a_k \bar{a}_l \rho^{|k-l|} e^{i(k-l)\varphi} .$$

This double series is absolutely convergent for $\rho \leq 1$, by virtue of the assumption of 5.12, and satisfies both 5.2 (1') and 5.2 (2'). There is no function different from (10) which could satisfy these conditions, since the solution of Dirichlet's problem is unique.

We observe that, in view of (7), the series

(11) $$F = \frac{1}{2} b_0 + \sum_{n=1}^{\infty} b_n \zeta^n ,$$

derived from (2) and (8), is also absolutely convergent.

[The foregoing can be used to prove another extremal property of the circle involving the stress function (see 5.2).

The stress function v *satisfies the inequality*

(12) $$2\pi v \leq A$$

where A *is the area of* D*. Equality is attained if, and only if,* D *is a circle and* v *is taken at its center*.

This is a particular case, concerned with two dimensions and stated in the language of the theory of torsion, of a more general theorem due to W. R. Wasow.*

As any point of D can be chosen as origin of the z-plane and image of the center of the unit circle in the ζ-plane, it is enough to prove (12) in assuming that $\zeta = 0$ and $a_0 = 0$. Under this assumption, we obtain by combining (7), (10) and 5.2 (3) that

$$2(v)_{z=0} = 2(v)_{\zeta=0} = b_0$$

$$= |a_1|^2 + |a_2|^2 + |a_3|^2 + \dots .$$

Yet, as is well known,

$$A = \pi (|a_1|^2 + 2|a_2|^2 + 3|a_3|^2 + \dots)$$

and, therefore,

$$\frac{2\pi(v)_{z=0}}{A} = \frac{|a_1|^2 + |a_2|^2 + |a_3|^2 + \dots}{|a_1|^2 + 2|a_2|^2 + 3|a_3|^2 + \dots} \leqq 1$$

with equality only if $a_2 = a_3 = \dots = 0$.]

5.14. Expressions for the torsional rigidity. By 5.2 (3) and 5.2 (6), we have

(1) $$P = \iint 2\Phi \, dx\, dy - \iint (x^2 + y^2)\, dx\, dy$$

$$= \int_0^1 \int_{-\pi}^{\pi} 2\Phi \left|\frac{dz}{d\zeta}\right|^2 \rho \, d\varphi \, d\rho - I_0 .$$

We have substituted polar coordinates in the ζ-plane for rectangular coordinates in the z-plane and used I_0 to denote the polar moment of inertia of D about the origin, in accordance with our notation. We shall need the expression

(2) $$I_0 = \int_0^1 \int_{-\pi}^{\pi} \left|z \frac{dz}{d\zeta}\right|^2 \rho \, d\varphi \, d\rho .$$

* Given in a lecture, Stanford University, June 8, 1950.

Using again 5.2 (3) and 5.2 (6), we can also write

(3) $$P = \iint [(x - \Phi_x)^2 + (y - \Phi_y)^2]\, dx\, dy$$

$$= \iint [x^2 - \Phi_x^2 + y^2 - \Phi_y^2]\, dx\, dy + 2E$$

where

(4) $$E = \iint (\Phi_x v_x + \Phi_y v_y)\, dx\, dy$$

$$= \iint [\partial(\Phi_x v)/\partial x + \partial(\Phi_y v)/\partial y]\, dx\, dy$$

$$= 0$$

by 5.2 (1') and 5.2 (2). From (3), (4) and 5.13 (1) we obtain

(5) $$P = I_0 - \iint (\Phi_x^2 + \Phi_y^2)\, dx\, dy$$

$$= I_0 - \iint \left|\frac{dF}{dz}\right|^2 dx\, dy$$

$$= I_0 - \int_0^1 \int_{-\pi}^{\pi} \left|\frac{dF}{d\zeta}\right|^2 \rho\, d\varphi\, d\rho .$$

We can express P in terms of the coefficients $a_0, a_1, a_2, \ldots$ of the series 5.12 (1), starting either from (1) or from (5). In both cases, however, we must compute I_0.

5.15. <u>Expansions of the polar moment of inertia</u>. By virtue of the assumption made in 5.12, all the series that we shall use in this section are absolutely convergent.

It follows from 5.12 (1) that

(1) $$z^2 = \sum_{n=0}^{\infty} A_n \zeta^n$$

where

(2) $$A_n = a_0 a_n + a_1 a_{n-1} + a_2 a_{n-2} + \cdots + a_n a_0 .$$

We obtain from (1) that

(3) $$2z \frac{dz}{d\zeta} = \sum_{n=0}^{\infty} n A_n \zeta^{n-1}$$

and now we can evaluate 5.14 (2):

(4) $$I_0 = (1/4) \int_0^1 \int_{-\pi}^{\pi} \sum_{k=0}^{\infty} kA_k \zeta^{k-1} \sum_{l=0}^{\infty} l\bar{A}_l \bar{\zeta}^{l-1} \rho \, d\varphi \, d\rho$$

$$= (\pi/2) \int_0^1 \sum_{n=0}^{\infty} n^2 A_n \bar{A}_n \rho^{2n-1} \, d\rho$$

$$= (\pi/4) \sum_{n=0}^{\infty} n|A_n|^2 .$$

We transform the last line of (4) as follows, supposing $|\zeta| = 1$ and using (1) and (3):

(5) $$I_0 = (1/8) \int_{-\pi}^{\pi} \sum_{k=0}^{\infty} k A_k \zeta^k \sum_{l=0}^{\infty} \bar{A}_l \zeta^{-l} \, d\varphi$$

$$= (1/8) \int_{-\pi}^{\pi} 2z \frac{dz}{d\zeta} \zeta \bar{z}^2 \, d\varphi$$

$$= (1/4) \int_{-\pi}^{\pi} z \bar{z} \, . \, \bar{z} \frac{dz}{d\zeta} \zeta . \, d\varphi$$

$$= (1/4) \int_{-\pi}^{\pi} \sum_{n=-\infty}^{\infty} b_n \zeta^n \sum_{m=-\infty}^{\infty} c_m \zeta^m \, d\varphi$$

$$= (\pi/2) \sum_{n=-\infty}^{\infty} b_n c_{-n} .$$

We have used 5.13 (4) and 5.13 (5) and introduced the notation

(6) $$\bar{z} \frac{dz}{d\zeta} \zeta = \sum_{l=0}^{\infty} \bar{a}_l \zeta^{-l} \sum_{k=0}^{\infty} k \dot{a}_k \zeta^k$$

$$= \sum_{n=-\infty}^{\infty} c_n \zeta^n ,$$

(7) $$c_n = \sum_{k-l=n} k\, a_k\, \bar{a}_l = \sum_{l=0}^{\infty} (n+1)\, a_{n+l}\, \bar{a}_l$$

in accordance with 5.13 (6).

16. <u>Expansions of the torsional rigidity</u>. We have obtained two expressions for P, 5.14 (1) and 5.14 (5), and two expressions for I_0, 5.15 (4) and 5.15 (5). We examine two different combinations of these formulae.

(a) We combine 5.14 (5) with 5.15 (5). Using 5.13 (11), we obtain

(1) $$P = (\pi/2) \sum_{n=-\infty}^{\infty} b_n c_{-n} - \int_0^1 \int_{-\pi}^{\pi} \sum_{k=0}^{\infty} k\, b_k \zeta^{k-1} \sum_{l=0}^{\infty} l\, \bar{b}_l \bar{\zeta}^{l-1} \rho\, d\varphi\, d\rho$$

$$= (\pi/2) \sum_{n=-\infty}^{\infty} b_n c_{-n} - \pi \sum_{n=0}^{\infty} n\, b_n \bar{b}_n$$

$$= (\pi/2) \sum_{n=-\infty}^{\infty} (b_n c_{-n} - |n|\, b_n b_{-n}) ;$$

we used also 5.13 (7).

(b) In combining 5.14 (1) with 5.15 (4), we shall find useful the following notation for fourfold sums:

$$\sum_{\alpha} \sum_{\beta} \sum_{\gamma} \sum_{\delta}$$

denotes a sum in which α, β, γ and δ run from 0 to ∞ independently of each other, but

$$\sum_{\alpha} \sum_{\beta} \sum_{\gamma} \sum_{\delta}{}^{*}$$

denotes a sum in which α, β, γ and δ take only those non-negative integral values which satisfy

(2) $$\alpha + \beta = \gamma + \delta .$$

Using this notation, we derive from 5.12 (1), 5.13 (10) and 5.14 (1) that

$$dz/d\zeta = e^{-i\varphi} \sum_{\beta=0}^{\infty} \beta\, a_\beta\, \rho^{\beta-1} e^{i\beta\varphi}$$

$$\overline{dz/d\zeta} = e^{i\varphi} \sum_{\delta=0}^{\infty} \delta \bar{a}_{\delta}\, \rho^{\delta-1} e^{-i\delta\varphi} ,$$

$$2\,\Phi = \sum_{\alpha=0}^{\infty} \sum_{\gamma=0}^{\infty} a_\alpha \bar{a}_\gamma\, \rho^{|\alpha-\gamma|} e^{i(\alpha-\gamma)\varphi} ,$$

$$(3)\quad P + I_0 = \int_0^1 \int_{-\pi}^{\pi} \sum_\alpha \sum_\beta \sum_\gamma \sum_\delta \beta\delta\, a_\alpha a_\beta \bar{a}_\gamma \bar{a}_\delta\, \rho^{|\alpha-\gamma|+\beta+\delta-1} e^{i(\alpha+\beta-\gamma-\delta)\varphi} d\varphi\, d\rho$$

$$= 2\pi \sum_\alpha \sum_\beta \sum_\gamma \sum_\delta{}^{*} a_\alpha a_\beta \bar{a}_\gamma \bar{a}_\delta\, \beta\delta/(|\alpha-\gamma| + \beta + \delta) .$$

On the other hand, we obtain from 5.15 (2) and 5.15 (4) that

$$(4)\quad I_0 = (\pi/4) \sum_\alpha \sum_\beta \sum_\gamma \sum_\delta{}^{*} (\alpha+\beta)\, a_\alpha a_\beta \bar{a}_\gamma \bar{a}_\delta .$$

We return now to (3), observe (2), distinguish the two cases $\alpha \geqq \gamma$ and $\alpha < \gamma$, and obtain

$$(5)\quad \begin{aligned} |\alpha-\gamma| + \beta + \delta &= 2 \max(\beta, \delta) , \\ 2\beta\delta/(|\alpha-\gamma| + \beta + \delta) &= \min(\beta, \delta) . \end{aligned}$$

Observe now that, interchanging α and β on one hand, and γ and δ on the other, we can write (3) in four different forms. Then we obtain from (3), (4) and (5)

$$P = (\pi/4) \sum_\alpha \sum_\beta \sum_\gamma \sum_\delta{}^{*} a_\alpha a_\beta \bar{a}_\gamma \bar{a}_\delta\, [\min(\alpha,\gamma) + \min(\alpha,\delta) + \min(\beta,\gamma) + \min(\beta,\delta) - \alpha - \beta]$$

and hence, after a short discussion in which we must use (2),

$$(6)\qquad P = (\pi/2) \sum_\alpha \sum_\beta \sum_\gamma \sum_\delta{}^{*} \min(\alpha,\beta,\gamma,\delta)\, a_\alpha a_\beta \bar{a}_\gamma \bar{a}_\delta .$$

We have derived two expressions for the torsional rigidity, (1) and (6). Expansion (1) has been first obtained by R. M. Morris who used a somewhat formal argument; see 16. A second derivation for (1), which exhibits more clearly the underlying conditions, is due to N. I. Muschelisvili.* Expansion (6) appears to be new. It is due to Professor H. Davenport who derived it by algebraic transformations from (1) and used it in

* See the exposition in Sokolnikoff's textbook, loc. cit., p. 170-176.

answering a question put by the authors that will be discussed in the next section.

5.17. Second proof of de Saint-Venant's theorem. We turn now to the following remarkable theorem:

> Of all cross-sections with a given area, the circular cross-section has the maximum torsional rigidity.*

This theorem has been clearly stated by de Saint-Venant who supported it by convincing physical considerations and several particular cases, but did not prove it in a mathematical sense (27). The general results on symmetrization which the authors of the present book have formerly obtained (24) yielded a first proof of the theorem (21). The expressions of P in terms of the coefficients of the mapping functions, derived in the foregoing section, open up a new access.

In fact, de Saint Venant's theorem can be expressed by the inequality

$$2\pi P \leq A^2 \tag{1}$$

where A denotes, as usual, the area of the cross-section. Yet, it is well known that A can be expressed in terms of the coefficients of the mapping function 5.12 (1):

$$A = \pi(|a_1|^2 + 2|a_2|^2 + 3|a_3|^2 + \dots) . \tag{2}$$

(In fact, in deriving 5.15 (4) we have followed the well known derivation of (2).) Combining (1) with (2) and recalling 5.7 (11), we obtain the double inequality

$$2\pi\left(\sum_{n=1}^{\infty} n|a_n|^2/(n+1)\right)^2 \leq P \leq (\pi/2)\left(\sum_{n=1}^{\infty} n|a_n|^2\right)^2 . \tag{3}$$

If we substitute for P its expression 5.16 (1), we obtain two inequalities between three series and we are led to the question: Can these inequalities be proved without reference to the fact that a_0, a_1, a_2, ... are the coefficients of a power series convergent and univalent in the unit circle ? (And a fortiori without reference to the physical interpretation of P.)

Professor Davenport gave an elegant answer to one half of this question of ours. He transformed 5.16 (1) into 5.16 (6) and applied to the latter

---

* We emphasize that only simply connected cross-sections are considered here. The case of multiply connected cross-sections, which presents new features, is considered in note A.

expression Cauchy's inequality

(4) $$|\sum u\bar{v}|^2 \leqq \sum |u|^2 \sum |v|^2 .$$

In fact, setting

(5) $$[\min(\alpha,\beta,\gamma,\delta)]^{1/2} a_\alpha a_\beta = u ,$$

$$[\min(\alpha,\beta,\gamma,\delta)]^{1/2} a_\gamma a_\delta = v ,$$

we derive from (4) that

(6) $$\sum_\alpha \sum_\beta \sum_\gamma \sum_\delta{}^{*} \min(\alpha,\beta,\gamma,\delta)\, a_\alpha a_\beta \bar{a}_\gamma \bar{a}_\delta$$

$$\leqq \sum_\alpha \sum_\beta \sum_\gamma \sum_\delta{}^{*} \min(\alpha,\beta,\gamma,\delta) |a_\alpha a_\beta|^2 .$$

We recollect 5.16 (2) and assume $\alpha \leqq \beta$. Then

$$\sum_{\gamma=0}^{\alpha+\beta} \min(\alpha,\beta,\gamma,\alpha+\beta-\gamma) = \sum_{\gamma=0}^{\alpha-1} \gamma + \sum_{\gamma=\alpha}^{\beta} \alpha + \sum_{\gamma=\beta+1}^{\alpha+\beta} (\alpha+\beta-\gamma)$$

$$= \alpha(\alpha-1)/2 + \alpha(\beta-\alpha+1) + \alpha(\alpha-1)/2$$

$$= \alpha\beta .$$

We obtain so from (6) and 5.16 (6) that

(7) $$2P/\pi = \sum_\alpha \sum_\beta \sum_\gamma \sum_\delta{}^{*} \min(\alpha,\beta,\gamma,\delta)\, a_\alpha a_\beta \bar{a}_\gamma \bar{a}_\delta$$

$$\leqq \sum_\alpha \sum_\beta \alpha\beta |a_\alpha a_\beta|^2$$

$$= \left(\sum_\alpha \alpha |a_\alpha|^2\right)^2 .$$

In view of (2), we have proved de Saint-Venant's theorem (1).

The great advantage of this proof is that we can discuss clearly the case of equality. Equality is attained in (4) if, and only if, the u are proportional to the v. In our case (5), the equality is attained if

$$a_\alpha a_\beta = a_\gamma a_\delta$$

whenever

$$\alpha + \beta = \gamma + \delta , \qquad \alpha, \beta, \gamma, \delta = 1, 2, 3, \ldots .$$

Especially

$$a_n \, a_1 = a_{n-1} \, a_2 \ .$$

Hence it follows that

$$a_n = \left(\frac{a_2}{a_1} \right)^{n-1} a_1 = q^{n-1} \, a_1 \qquad \text{for } n = 2, \, 3, \, 4, \ldots,$$

(8) $$z = a_0 + \frac{a_1 \, \zeta}{1 - q \, \zeta}.$$

by 5.12 (1) (q = 0 is permissible). Now, the function (8) maps the circle $|\zeta| < 1$ onto a circle of the z-plane. That is, only in the case of the circle is equality attained in (7). The circular cross-section has definitely greater torsional rigidity than any other cross-section with equal area.

5.18. Relaxing the assumption. If the series 5.12 (1) converges in the circle $|\zeta| < 1$ and represents there a univalent function, the assumption of 5.12 need not be satisfied. Yet this assumption is certainly satisfied for the series

(1) $$z = a_0 + a_1 \, \tau \; \zeta + a_2 \, \tau^2 \, \zeta^2 + \ldots + a_n \, \tau^n \, \zeta^n + \ldots$$

where τ is a fixed real number, $0 < \tau < 1$. This series maps the circle $|\zeta| < 1$ onto a domain $\underline{D}_\tau$ contained in $\underline{D}$. Working with $\underline{D}_\tau$ and (1), instead of $\underline{D}$ and 5.12 (1), we could derive the expansions 5.16 (1) and 5.16 (6) for the torsional rigidity of $\underline{D}_\tau$. Finally, using inequality 5.17 (7), we could pass to the limit, making τ tend to 1, and derive so the expression 5.16 (6) for the torsional rigidity P of $\underline{D}$ itself, under the new assumption that the area of $\underline{D}$ is finite. This assumption amounts to the same as convergence of

(2) $$|a_1|^2 + 2|a_2|^2 + \ldots + n|a_n|^2 + \ldots$$

and is certainly less restrictive than the one we made before, namely that 5.12 (2) converges. We could not use the same method to derive the older expansion 5.16 (1) for the torsional rigidity; this latter could possibly diverge even if (2) converges.

The reasoning just sketched admits a larger class of domains and extends the scope of de Saint Venant's theorem.

5.19. Another expansion of the polar moment of inertia. We generalize the considerations of section 5.15. We assume that the series

(1)
$$z = \sum_{n=-\infty}^{\infty} a_n \zeta^n .$$

and its formal, term-by-term derivative are absolutely convergent on the circle $|\zeta| = 1$ and that (1) establishes a one-to-one correspondence between the points of the circle $|\zeta| = 1$ and a closed curve *C* without double points in the z-plane. Our aim is to compute I_0, the polar moment of inertia of the domain *D* bounded by *C* about the origin. We shall vary slightly the method of section 5.15 where we dealt with the particular case 5.12 (1) of our present series (1), which corresponds to the restriction 5.13 (6).

We set

(2)
$$z = r\, e^{i\theta} , \qquad \zeta = \rho\, e^{i\varphi} ,$$

r, θ, ρ, φ real, $r \geq 0$, $\rho \geq 0$. Then,

(3)
$$4\, I_0 = \int_{-\pi}^{\pi} r^4\, d\theta .$$

We assume here that $z = r\, e^{i\theta}$, linked to ζ by (1), describes *C* in the positive direction when ζ so describes the circle $|\zeta| = 1$; in the other case the sign on one side of (3) should be changed. Observing that

$$dz = e^{i\theta}\, (dr + i\, r\, d\theta) ,$$

$$\bar{z}\, dz = r\, dr + i\, r^2\, d\theta ,$$

we rewrite (3) in the form

(4)
$$4\, I_0 = \int_{C} z\, \bar{z}\, \Im\, \bar{z}\, dz$$

$$= \Im \int_{-\pi}^{\pi} \bar{z}^2\, z\, (dz/d\zeta)\, i\zeta\, d\varphi$$

$$= \Re \int_{-\pi}^{\pi} \bar{z}^2\, z\, (dz/d\zeta)\zeta\, d\varphi .$$

Setting

(5)
$$z^2 = \sum_{n=-\infty}^{\infty} A_n\, \zeta^n$$

we transform (4) into

(6) $$4\, I_0 = (1/2) \int_{-\pi}^{\pi} \sum_k \bar{A}_k \zeta^{-k} \sum_l l A_l \zeta^l \, d\varphi$$

$$= \pi \sum_{n=-\infty}^{\infty} n |A_n|^2 .$$

This goes over into 5.15 (4) under the condition 5.13 (6).

We are mainly interested in the particular case in which the series (1) takes the form

(7) $$z = a_1 \zeta + a_0 + a_{-1} \zeta^{-1} + a_{-2} \zeta^{-2} + \ldots$$

and establishes a one-to-one conformal mapping of $|\zeta| > 1$ onto the exterior of the curve C. In this case

(8) $$|a_1| = \bar{r}$$

the outer radius of D,

(9) $$A_3 = A_4 = A_5 = \ldots = 0 ,$$

$$A_2 = a_1^2 , \qquad A_1 = 2\, a_0\, a_1 , \qquad A_0 = a_0^2 + 2\, a_1\, a_{-1}, \ldots$$

and (6) reduces to

(10) $$4\, I_0 = \pi \left(2\, |a_1|^4 + 4\, |a_0\, a_1|^2 - |A_{-1}|^2 - 2\, |A_{-2}|^2 - \ldots \right)$$

We wish now to estimate I, the polar moment of inertia of D with respect to its center of gravity. We know that

(11) $$I \leqq I_0 .$$

Moreover, I remains unchanged (I_0 would not remain so) when the domain undergoes a translation. Therefore, we may substitute $z - a_0$ for z, or, what amounts to the same, we may assume that $a_0 = 0$. Thus we obtain from (8), (10) and (11)

(12) $$4\, I \leqq \pi\, (2\bar{r}^4 - |A_{-1}|^2 - 2|A_{-2}|^2 - \ldots) ,$$

$$2\, I < \pi\, \bar{r}^4$$

unless

$$A_{-1} = A_{-2} = A_{-3} = \ldots = 0 ,$$

that is, unless

$$(z - a_0)^2 = A_2 \zeta^2 + A_0$$

in which case the equation of the curve *C* in the z-plane reduces to

$$|(z - a_0)^2 - A_0| = |A_2| \quad . \tag{13}$$

That is, *C* *is a level curve of the absolute value of a polynomial of the second degree*, or a "lemniscate of the second order" as such a level curve is called occasionally. We state our result with this terminology: *For an arbitrary plane domain* *D*

$$I \leqq \pi \bar{r}^4/2$$

and equality is attained if, and only if, the boundary of *D* *is a lemniscate of the second order*. A familiar reasoning (similar to that sketched in 5.18) shows that this result is free from any restriction concerning the convergence of the series (7) on the circle $|\zeta| = 1$.

CHAPTER VI. NEARLY CIRCULAR AND NEARLY SPHERICAL DOMAINS

6.1. Notation. In the following sections we use the notation of 1.32 - 1.35, that is, we consider the domain $r \leqq 1 + \rho(\varphi)$ in the plane and the domain $r \leqq 1 + \rho(\theta, \varphi)$ in space. The coefficients of the Fourier series

(1) $$\rho(\varphi) = a_0 + 2 \sum_{n=1}^{\infty} (a_n \cos n\varphi + b_n \sin n\varphi)$$

are of the first order, and the same holds for the terms of the Laplace series

(2) $$\rho(\theta, \varphi) = X_0(\theta, \varphi) + X_1(\theta, \varphi) + \dots + X_n(\theta, \varphi) + \dots .$$

All the formulas obtained will involve only terms up to the second order inclusively, the terms of higher than second order being neglected. For instance, the expressions for the area-radius and the perimeter-radius of a nearly circular plane domain,

(3) $$\bar{A} = 1 + a_0 + \sum_{n=1}^{\infty} (a_n^2 + b_n^2)$$

and

(4) $$\bar{L} = 1 + a_0 + \sum_{n=1}^{\infty} n^2 (a_n^2 + b_n^2)$$

have the same initial term 1 and the same term of first order a_0. The second order terms, represented by infinite sums, are different, and terms of higher order are omitted.

The derivation of (3) and (4) by the method of the next section presents no difficulty and can be left to the reader.

6.2. Moment of inertia. We begin by computing the centroid $c = |c|e^{i\gamma}$ of a uniform distribution of matter over the area $r \leqq 1 + \rho$. Neglecting terms of higher than the first order, we have, as $A - \pi$ is of the first order,

$$c = \pi^{-1} \int_{-\pi}^{\pi} \int_{0}^{1+\rho(\varphi)} r\, e^{i\varphi} \cdot r\, d\,r\, d\varphi$$

$$= (3\pi)^{-1} \int_{-\pi}^{\pi} (1+\rho)^3 e^{i\varphi}\, d\varphi = (3\pi)^{-1} \int_{-\pi}^{\pi} (1+3\rho)\, e^{i\varphi}\, d\varphi$$

and so*

(1) $$c = 2\,(a_1 + ib_1)\ .$$

Then we compute the moment of inertia with respect to the origin, retaining the terms of second order:

(2) $$I_0 = (1/4) \int_{-\pi}^{\pi} r^4\, d\varphi = (1/4) \int_{-\pi}^{\pi} (1 + 4\rho + 6\rho^2)\, d\varphi$$

$$= \pi/2 + 2\pi a_0 + 3\pi\,[a_0^2 + 2 \sum_{n=1}^{\infty} (a_n^2 + b_n^2)]\ .$$

We obtain from (1) and (2) that

(3) $$I_c = I_0 - A|c|^2 = I_0 - 4\pi(a_1^2 + b_1^2)$$

$$= (\pi/2)[1 + 4a_0 + 6a_0^2 + 4(a_1^2 + b_1^2) + 12 \sum_{n=2}^{\infty} (a_n^2 + b_n^2)]\ .$$

Finally

(4) $$\bar{I}_c = (2\, I_c/\pi)^{1/4}$$

$$= 1 + a_0 + a_1^2 + b_1^2 + 3 \sum_{n=2}^{\infty} (a_n^2 + b_n^2)$$

which agrees with no. 3 of the table in 1.33.

We note for comparison that

(5) $$\bar{I}_0 = (2\, I_0/\pi)^{1/4} = 1 + a_0 + 3 \sum_{n=1}^{\infty} (a_n^2 + b_n^2)\ .$$

6.3. Outer and inner radius.** (a) We seek Green's function G of the

---

* The result would be the same (except for terms of the second and higher order) for a uniform distribution over the curve $r = 1 + \rho$.

** In this section and in other sections of the present chapter dealing with boundary value problems, the treatment is essentially heuristic. The transition to a more complete treatment is, of course, easier in the present section than in some of the following cases.

exterior domain $r > 1 + \rho$ in the form

(1) $$G = \log r - p(r, \varphi) - q(r, \varphi) ,$$

(2) $$p(r, \varphi) = \sum_{n=0}^{\infty} (A_n \cos n\varphi + B_n \sin n\varphi) r^{-n} ,$$

$$q(r, \varphi) = \sum_{n=0}^{\infty} (A_n' \cos n\varphi + B_n' \sin n\varphi) r^{-n} .$$

Here A_n, B_n are quantities of the first order while A_n', B_n' are of the second order. The outer radius is obtained from

(3) $$\log \bar{r} = A_0 + A_0' , \qquad \bar{r} = 1 + A_0 + A_0' + A_0^2/2 .$$

On the curve $r = 1 + \rho$ we have $G = 0$ so that

(4) $$\rho - \rho^2/2 - p(1, \varphi) - \rho p_r(1, \varphi) - q(1, \varphi) = 0 .$$

The terms of the first order yield

$$\rho(\varphi) - p(1, \varphi) = 0$$

so that

(5) $$p(r, \varphi) = a_0 + 2 \sum_{n=1}^{\infty} (a_n \cos n\varphi + b_n \sin n\varphi) r^{-n} .$$

In particular

(6) $$A_0 = a_0 .$$

Taking the mean value of the second order terms in (4) we obtain

$$-a_0^2/2 - \sum_{n=1}^{\infty} (a_n^2 + b_n^2) + \sum_{n=1}^{\infty} 2n (a_n^2 + b_n^2) - A_0' = 0 .$$

Combining this with (3) and (6), we obtain

(7) $$\bar{r} = 1 + a_0 + \sum_{n=1}^{\infty} (2n - 1) (a_n^2 + b_n^2)$$

which justifies 1.33 no. 2.

(b) We seek now Green's function G_c of the interior domain with respect to the centroid c; see 6.2 (1). Our notation is similar to that used in (a), but has, of course, a different meaning:

(8) $$G_c = \log r' - p(r, \varphi) - q(r, \varphi) ,$$

(9) $$p(r, \varphi) = \sum_{n=0}^{\infty} (A_n \cos n\varphi + B_n \sin n\varphi)\, r^n ,$$

$$q(r, \varphi) = \sum_{n=0}^{\infty} (A'_n \cos n\varphi + B'_n \sin n\varphi)\, r^n .$$

Here r' stands for the distance of the point $re^{i\varphi}$ from the centroid $c = |c|\, e^{i\gamma}$. The coefficients of p are of the first order, those of q of the second order. The inner radius r_c is determined by

$$\log r_c = p(|c|, \gamma) + q(|c|, \gamma)$$

$$= A_0 + (A_1 \cos \gamma + B_1 \sin \gamma)\, |c| + A'_0 ,$$

(10) $$r_c = 1 + A_0 + (A_1 \cos \gamma + B_1 \sin \gamma)\, |c| + A'_0 + A_0^2/2 .$$

On the boundary $G_c = 0$ so that (8) yields

(11) $$\log r' = (1/2) \log [|c|^2 + r^2 - 2|c|r \cos (\varphi - \gamma)]$$

$$= \log r - |c|r^{-1} \cos (\varphi - \gamma) - (1/2)|c|^2 r^{-2} \cos 2(\varphi - \gamma)$$

$$= \rho - \tfrac{1}{2}\rho^2 - |c|(1 - \rho) \cos (\varphi - \gamma) - \tfrac{1}{2} |c|^2 \cos 2(\varphi - \gamma)$$

$$= p(1, \varphi) + \rho\, p_r (1, \varphi) + q(1, \varphi) .$$

Comparing the terms of the first order in the last two lines of (11), we find that

(12) $$\rho - |c| \cos (\varphi - \gamma) = p(1, \varphi)$$

so that, in view of 6.2 (1),

(13) $$p(r, \varphi) = a_0 + 2 \sum_{n=2}^{\infty} (a_n \cos n\varphi + b_n \sin n\varphi)\, r^n ,$$

in particular

(14) $$A_0 = a_0 , \qquad A_1 = B_1 = 0 .$$

Observe also that, by 6.2 (1),

(15) $$a_1 \cos \gamma + b_1 \sin \gamma = 2 (a_1^2 + b_1^2)\, |c|^{-1} .$$

Comparing the mean values of the terms of the second order in the last two lines of (11), and using (15), we find that

(16) $$-\tfrac{1}{2} a_0^2 - \sum_{n=1}^{\infty} (a_n^2 + b_n^2) + 2 (a_1^2 + b_1^2) = \sum_{n=2}^{\infty} 2n(a_n^2 + b_n^2) + A'_0 .$$

From (10), (14) and (16) we obtain finally

$$(17) \qquad r_c = 1 + a_0 + a_1^2 + b_1^2 - \sum_{n=2}^{\infty} (2n + 1)(a_n^2 + b_n^2)$$

which justifies 1.33 no. 7.

We note for comparison that the inner radius r_0 of the nearly circular domain with respect to the origin is given by the expansion

$$(18) \qquad r_0 = 1 + a_0 - \sum_{n=1}^{\infty} (2n + 1)(a_n^2 + b_n^2) .$$

6.4. Torsional rigidity. As we have discussed in 5.2, the differential equation of this problem is

$$(1) \qquad \nabla^2 v + 2 = 0$$

with the boundary condition $v = 0$. The torsional rigidity is given by

$$(2) \qquad P = 2 \iint v \, dx \, dy .$$

The domain of integration in (2) is the interior of $\underline{C}$.

In dealing with a nearly circular domain, we set

$$\begin{aligned} (3) \qquad 2v &= 1 - r^2 + p(r, \varphi) + q(r, \varphi) \\ &= 1 - r^2 + \sum_{n=0}^{\infty} (A_n \cos n\varphi + B_n \sin n\varphi) r^n \\ &\quad + \sum_{n=0}^{\infty} (A_n' \cos n\varphi + B_n' \sin n\varphi) r^n . \end{aligned}$$

The functions $p(r, \varphi)$ and $q(r, \varphi)$ are harmonic, the coefficients A_n, B_n of p are of the first order and the coefficients A_n', B_n' of q of the second order; (3) satisfies (1). By (2) and (3)

$$\begin{aligned} (4) \quad P &= A - I_0 + \sum_{n=0}^{\infty} \int_{-\pi}^{\pi} (A_n \cos n\varphi + B_n \sin n\varphi) \frac{(1 + \rho)^{n+2}}{n + 2} d\varphi + \pi A_0' \\ &= A - I_0 + \pi A_0 + \sum_{n=0}^{\infty} \int_{-\pi}^{\pi} (A_n \cos n\varphi + B_n \sin n\varphi) \rho d\varphi + \pi A_0' ; \end{aligned}$$

A denotes the area and I_0 the moment of inertia about the origin, as usual.

On the boundary $r = 1 + \rho$ we have

$$(5) \qquad -2\rho - \rho^2 + p(1, \varphi) + p_r(1, \varphi)\rho + q(1, \varphi) = 0.$$

The terms of the first order yield $-2\rho + p(1, \varphi) = 0$; thus

(6) $$p(r, \varphi) = 2\, a_0 + 4 \sum_{n=1}^{\infty} (a_n \cos n\varphi + b_n \sin n\varphi)\, r^n .$$

Taking the mean value of the second order terms in (5), we obtain

(7) $$- a_0^2 - 2 \sum_{n=1}^{\infty} (a_n^2 + b_n^2) + \sum_{n=1}^{\infty} 4n(a_n^2 + b_n^2) + A_0' = 0 .$$

In view of 6.1 (3), 6.2 (2), (6) and (7), we obtain from (4)

(8) $$P/\pi = \frac{1}{2} + 2\, a_0 + 3\, a_0^2 - 2 \sum_{n=1}^{\infty} (2n - 3)\, (a_n^2 + b_n^2) .$$

Since $P = \pi \bar{P}^4/2$, we finally find that

(9) $$\bar{P} = 1 + a_0 - \sum_{n=1}^{\infty} (2n - 3)\, (a_n^2 + b_n^2)$$

which agrees with 1.33 no. 5.

6.5. <u>The principal frequency of a nearly circular membrane</u>. (a) As we have discussed in 5.3, the differential equation of this problem is:

(1) $$\nabla^2 w + \lambda^2 w = 0$$

with the boundary condition $w = 0$. This equation admits the solutions $J_n(\lambda r)\, e^{in\varphi}$, n an integer. In the case of the unit circle $r \leqq 1$ the first characteristic function is $w(r) = J_0(jr)$ where j has the usual meaning; see 1.3 (1).

For the domain $r \leqq 1 + \rho$ we seek the solution of this problem in the form

(2) $$w = w(r, \varphi) = J_0(\lambda r) - p(\lambda r, \varphi) - q(\lambda r, \varphi) ,$$

(3) $$p(r, \varphi) = \sum_{n=0}^{\infty} J_n(r)\, (A_n \cos n\varphi + B_n \sin n\varphi),$$
$$q(r, \varphi) = \sum_{n=0}^{\infty} J_n(r)\, (A_n' \cos n\varphi + B_n' \sin n\varphi),$$

(4) $$\lambda = \Lambda = \Lambda_1 = j + l + l'$$

where the quantities A_n, B_n, l are of the first and the quantities A_n', B_n', l' of the second order. Substituting $r = 1 + \rho$, and using the boundary condition, we find

$$\lambda r = j + (l + j\rho) + (l' + l\rho),$$

$$(5) \qquad w = J_0(j) + (l + j\rho + l' + l\rho)\, J_0'(j) + \tfrac{1}{2}(l + j\rho)^2\, J_0''(j)$$
$$- p(j, \varphi) - (l + j\rho)\, p_r(j, \varphi) - q(j, \varphi) = 0 .$$

The first order terms in (5) yield:

$$(l + j\rho)\, J_0'(j) - p(j, \varphi) = 0 ,$$

or

$$(6) \qquad p(r, \varphi) = A_0 J_0(r) + 2j\, J_0'(j) \sum_{n=1}^{\infty} \frac{J_n(r)}{J_n(j)} (a_n \cos n\varphi + b_n \sin n\varphi),$$

$$(7) \qquad l = -ja_0 .$$

Taking the mean value of the second order terms in (5), using (6) and (7), and observing that $jJ_0''(j) = -J_0'(j)$, we obtain

$$(8) \qquad l' = ja_0^2 + j \sum_{n=1}^{\infty} \left(1 + \frac{2j\, J_n'(j)}{J_n(j)}\right) (a_n^2 + b_n^2) .$$

Since $\bar{\Lambda} = j/\Lambda$, (4), (7) and (8) yield 1.32 (3).

(b) We shall show that

$$(9) \qquad -2n - 1 < -1 - \frac{2j\, J_n'(j)}{J_n(j)} \leqq -2n + 3$$

<u>for</u> $n \geq 1$, <u>and that the case of equality is attained only for</u> $n = 1$. This will justify the place of $\bar{\Lambda} = \bar{\Lambda}_1$ in the table of 1.33, between r_c and $\bar{P}$.

The case $n = 1$ follows from the differential equation of $J_0(x)$, since $J_1(x) = -J_0'(x)$. Let $j_{n1}, j_{n2}, j_{n3}, \ldots$ denote the positive zeros of $J_n(x)$,

$$(10) \qquad j_{n1} < j_{n2} < \cdots < j_{n\nu} < \cdots .$$

We have*

$$(11) \qquad \frac{J_n'(x)}{J_n(x)} = \frac{n}{x} + \sum_{\nu=1}^{\infty} \frac{-2x}{j_{n\nu}^2 - x^2}$$

* Cf. G. N. Watson, A treatise on the theory of Bessel functions, p. 498 (3).

so that

(12) $$-1 - \frac{2j\, J_n'(j)}{J_n(j)} = -1 - 2n + \sum_{\nu=1}^{\infty} \frac{4\, j^2}{j_{n\nu}^2 - j^2} .$$

The zeros $j_{n\nu}$ are increasing if ν is fixed and n is increasing.* Thus each term of the sum on the right hand side of (12) is positive for $n \geq 1$ which makes the lower bound (9) obvious. Moreover the same terms decrease as n increases so that the largest value for the sum arises if n = 1 in which case the right-hand inequality (9) becomes an equation, according to the remark made above.

This establishes the assertion.

6.6. The clamped plate. (a) The differential equation of this problem is

(1) $$\nabla^4 u - \lambda^4 u = 0$$

with the boundary conditions

(2) $$u = 0, \qquad \partial u/\partial n = 0 \qquad \text{for } r = 1 + \rho .$$

The condition $\partial u/\partial n = 0$ can be replaced by $\partial u/\partial r = 0$. Equation (1) admits the solutions

(3) $$J_n(\lambda r)\, e^{in\varphi}, \qquad I_n(\lambda r)\, e^{in\varphi} ;$$

n is an integer. For the unit circle r = 1 the first characteristic function is v (hr), where

(4) $$v(r) = J_0(r)\, I_0'(h) - J_0'(h)\, I_0(r)$$

and

(5) $$h = 3.1962$$

is the smallest positive root of the equation v(h) = 0; see e.g. 26, vol. 1, pp. 366 - 367.

We seek u in the form

(6) $$u = u(r, \varphi) = v(\lambda r) - p(\lambda r, \varphi) - q(\lambda r, \varphi),$$

(7) $$p(r,\varphi) = \sum_{n=0}^{\infty} J_n(r)(A_n \cos n\varphi + B_n \sin n\varphi) + \sum_{n=0}^{\infty} I_n(r)(C_n \cos n\varphi + D_n \sin n\varphi),$$

---

* See, Watson, loc. cit., p. 508 (2).

$$q(r,\varphi) = \sum_{n=0}^{\infty} J_n(r)(A_n' \cos n\varphi + B_n' \sin n\varphi) + \sum_{n=0}^{\infty} I_n(r)(C_n' \cos n\varphi + D_n' \sin n\varphi) ,$$

$$\text{(8)} \qquad \lambda = \Lambda_2 = h + l + l' ,$$

where A_n, B_n, C_n, D_n, l are of the first order and A_n', B_n', C_n', D_n', l' of the second order. Substituting $r = 1 + \rho$, we find

$$\text{(9)} \qquad \lambda r = h + (l + h\rho) + (l' + l\rho)$$

and express the boundary conditions as follows:

$$u = v(h) + (l + h\rho + l' + l\rho)\, v'(h) + \tfrac{1}{2}(l + h\rho)^2\, v''(h)$$

$$\text{(10)} \qquad - p(h,\varphi) - (l + h\rho)\, p_r(h,\varphi) - q(h,\varphi) = 0,$$

$$\lambda^{-1} \cdot \frac{\partial u}{\partial r} = v'(h) + (l + h\rho + l' + l\rho)\, v''(h) + \tfrac{1}{2}(l + h\rho)^2\, v'''(h)$$

$$\text{(11)} \qquad - p_r(h,\varphi) - (l + h\rho)\, p_{rr}(h,\varphi) - q_r(h,\varphi) = 0 .$$

In view of the fact that $v(h) = v'(h) = 0$ the first order terms yield

$$\text{(12)} \qquad \begin{aligned} p(h,\varphi) &= 0 , \\ (l + h\rho)\, v''(h) - p_r(h,\varphi) &= 0 . \end{aligned}$$

Since $v''(h) \neq 0$, these two conditions are equivalent to the following conditions:

$$\text{(13)} \qquad \begin{aligned} &p(r,\varphi) = kw_0(r) + 2h\, v''(h) \sum_{n=1}^{\infty} \frac{w_n(r)}{w_n'(h)} (a_n \cos n\varphi + b_n \sin n\varphi) , \\ &w_n(r) = J_n(r)\, I_n(h) - I_n(r)\, J_n(h), \qquad l = - h\, a_0 \end{aligned}$$

where k is an arbitrary constant.

Now we form the mean values of the second order terms in (10) and (11):

$$-h^2 v''(h) \sum_{n=1}^{\infty} (a_n^2 + b_n^2) - (J_0(h)\, A_0' + I_0(h)\, C_0') = 0 ,$$

$$\text{(14)} \qquad (l' - h a_0^2) v''(h) + h^2 v'''(h) \sum_{n=1}^{\infty} (a_n^2 + b_n^2) - 2h^2 v''(h) \sum_{n=1}^{\infty} \frac{w_n''(h)}{w_n'(h)} (a_n^2 + b_n^2)$$

$$-(J_0'(h)\, A_0' + I_0'(h)\, C_0') = 0 .$$

Taking into account the relation $v(h) = 0$, we find that

$$J_0'(h)\, A_0' + I_0'(h)\, C_0' = \frac{J_0'(h)}{J_0(h)} (J_0(h)\, A_0' + I_0(h)\, C_0')$$

$$= - h^2 v''(h) \frac{J_0'(h)}{J_0(h)} \sum_{n=1}^{\infty} (a_n^2 + b_n^2)$$

and so we conclude from (14) that

$$(15) \qquad l' = h\, a_0^2 - h^2 \frac{v'''(h)}{v''(h)} \sum_{n=1}^{\infty} (a_n^2 + b_n^2) + 2h^2 \sum_{n=1}^{\infty} \frac{w_n''(h)}{w_n'(h)} (a_n^2 + b_n^2)$$

$$- h^2 \frac{J_0'(h)}{J_0(h)} \sum_{n=1}^{\infty} (a_n^2 + b_n^2) .$$

We find from the differential equation of the Bessel functions and the definitions of v and w_n that

$$- \frac{v'''(h)}{v''(h)} = \frac{1}{h} - \frac{J_0'(h)}{J_0(h)} ,$$

$$\frac{w_n''(h)}{w_n'(h)} = - \frac{1}{h} + 2 \left(\frac{I_n'(h)}{I_n(h)} - \frac{J_n'(h)}{J_n(h)} \right)^{-1}$$

and so we conclude from (8), (13) and (15) that

$$(16) \quad \Lambda_2 = h - ha_0 + ha_0^2 + \sum_{n=1}^{\infty} \left\{ -h - \frac{2h^2 J_0'(h)}{J_0(h)} + 4h^2 \left(\frac{I_n'(h)}{I_n(h)} - \frac{J_n'(h)}{J_n(h)} \right)^{-1} \right\} (a_n^2 + b_n^2).$$

Finally $\bar{\Lambda}_2 = h/\Lambda_2$, and hence no. 8 of the table 1.33 can be easily justified.

(b) As we have just proved, the coefficient of $a_1^2 + b_1^2$ in the expression of $\bar{\Lambda}_2$ is:

$$1 + \frac{2h\, J_0'(h)}{J_0(h)} - 4h \left(\frac{I_1'(h)}{I_1(h)} - \frac{J_1'(h)}{J_1(h)} \right)^{-1}$$

$$= 1 + \frac{2h\, J_0'(h)}{J_0(h)} - 4h \left(\frac{I_0''(h)}{I_0'(h)} - \frac{J_0''(h)}{J_0'(h)} \right)^{-1}$$

In view of Bessel's differential equation and the definition of h we obtain

$$\frac{I_0''(h)}{I_0'(h)} - \frac{J_0''(h)}{J_0'(h)} = \frac{I_0(h)}{I_0'(h)} + \frac{J_0(h)}{J_0'(h)} = \frac{2\, J_0(h)}{J_0'(h)}$$

so that the coefficient in question is = 1, in accordance with 1.33 (5).

(c) The inequality

$$1 + \frac{2h\, J_0'(h)}{J_0(h)} - 4h\left(\frac{I_n'(h)}{I_n(h)} - \frac{J_n'(h)}{J_n(h)}\right)^{-1} < -2n - 1 \tag{17}$$

is false for n = 2 and correct for $n \geq 3$.

Indeed,

$$1 + \frac{2h\, J_0'(h)}{J_0(h)} = 6.27 \,, \tag{18}$$

$$4h\left(\frac{I_2'(h)}{I_2(h)} - \frac{J_2'(h)}{J_2(h)}\right)^{-1} = 11.06 \tag{19}$$

and $-4.79 > -5$.

On the other hand we conclude from 6.5 (11) that for $n \geq 3$

$$\begin{aligned} \frac{I_n'(h)}{I_n(h)} - \frac{J_n'(h)}{J_n(h)} &= \sum_{\nu=1}^{\infty} \left(\frac{2h}{j_{n\nu}^2 + h^2} + \frac{2h}{j_{n\nu}^2 - h^2}\right) \\ &= 4h \sum_{\nu=1}^{\infty} j_{n\nu}^{-2}\, [1 - (h/j_{n\nu})^4]^{-1} \\ &< 4h\, [1 - (h/j_{31})^4]^{-1} \sum_{\nu=1}^{\infty} j_{n\nu}^{-2} \\ &= h[1 - (h/j_{31})^4]^{-1} \frac{1}{n+1} \end{aligned} \tag{20}$$

(cf. Watson, loc. cit., p. 502). Therefore, the left hand side of (17) is less than

$$\begin{aligned} &6.27 - 4(n + 1)\,(1 - (h/j_{31})^4) \\ = \;&6.27 - 3.75\,(n + 1) < -2n - 1 \end{aligned}$$

for $n \geq 3$, and so (17) follows.

(d) The result under (c), combined with 6.5 (b), shows for $n \geq 3$ that the coefficient of $a_n^2 + b_n^2$ is greater in the expansion of $\overline{\Lambda}_1$ than in that of $\overline{\Lambda}_2$. This, however, remains true also in the case n = 2:

$$-2.78 > -4.79 \,.$$

[6.6 A. <u>Buckling of a plate</u>. (a) The differential equation of this problem is

$$\nabla^4 u + \lambda^2 \nabla^2 u = 0 \tag{1}$$

with the boundary conditions

(2) $$u = 0, \qquad \partial u/\partial n = 0 \qquad \text{for } r = 1 + \rho .$$

Again we replace $\partial u/\partial n = 0$ by the condition $\partial u/\partial r = 0$. Equation (1) admits the solutions of 6.5 (1) and of $\nabla^2 u = 0$, i.e., the functions

(3) $$J_n(\lambda r)e^{in\varphi}, \qquad r^n e^{in\varphi}, \qquad n \text{ integer}.$$

For the unit circle $r = 1$ the first eigenfunction is $v(kr)$ where

(4) $$v(r) = J_0(r) - J_0(k)$$

and

(5) $$k = 3.8317$$

is the smallest positive root of the equation $v'(k) = 0$.

The further argument is almost exactly the same as in 6.6 so that we point out only the necessary modifications. We seek u in the form

(6) $$u = u(r, \varphi) = v(\lambda r) - p(\lambda r, \varphi) - q(\lambda r, \varphi),$$

(7) $$\begin{cases} p(r,\varphi) = \sum_{n=0}^{\infty} J_n(r)(A_n \cos n\varphi + B_n \sin n\varphi) + \sum_{n=0}^{\infty} r^n(C_n \cos n\varphi + D_n \sin n\varphi), \\ q(r,\varphi) = \sum_{n=0}^{\infty} J_n(r)(A_n' \cos n\varphi + B_n' \sin n\varphi) + \sum_{n=0}^{\infty} r^n(C_n' \cos n\varphi + D_n' \sin n\varphi), \end{cases}$$

(8) $$\lambda = \Lambda_3 = k + l + l'$$

where the constants have the same order as in 6.6 (7), (8). No change is necessary in 6.6 (9) - (13) except that in the last formula $w_n(r)$ has the following meaning:

(9) $$w_n(r) = J_n(r)\, k^n - r^n\, J_n(k),$$

i.e., $I_n(r)$ is replaced by r^n. Also h has to be replaced by k. Now we can accept 6.6 (14) and (15) with the change that h must be replaced by k, $I_0(h)$ by 1, $I_0'(h)$ and $J_0'(h)$ by 0.

Thus we obtain

(10) $$\begin{cases} l = -ka_0, \\ l' = ka_0^2 - k^2 \frac{v'''(k)}{v''(k)} \sum_{n=1}^{\infty} (a_n^2 + b_n^2) + 2k^2 \sum_{n=1}^{\infty} \frac{w_n''(k)}{w_n'(k)} (a_n^2 + b_n^2) \end{cases}$$

where $v(r)$ and $w_n(r)$ have the meaning (4) and (9), respectively. Now

$$-\frac{v'''(k)}{v''(k)} = -\frac{J_0'''(k)}{J_0''(k)} = \frac{1}{k} ,$$

$$\frac{w_n''(k)}{w_n'(k)} = -\frac{1}{k} + \left(\frac{n}{k} - \frac{J_n'(k)}{J_n(k)}\right)^{-1}$$

so that from (8) and (10) we conclude

$$(11) \qquad \Lambda_3 = k - ka_0 + ka_0^2 + \sum_{n=1}^{\infty} \left\{ - k + 2k^2 \left(\frac{n}{k} - \frac{J_n'(k)}{J_n(k)}\right)^{-1} \right\} (a_n^2 + b_n^2) .$$

Finally $\bar{\Lambda}_3 = k/\Lambda_3$ and no. 9 of the table 1.33 follows.

We note the following alternate expression for the coefficient of $a_n^2 + b_n^2$ in the final result:

$$(12) \qquad 1 - 2k\left(\frac{n}{k} - \frac{J_n'(k)}{J_n(k)}\right)^{-1} = 1 - 2k\,\frac{J_n(k)}{J_{n+1}(k)}$$

For $n = 1$ this is obviously $= 1$.

(b) We prove the inequality

$$(13) \qquad \begin{aligned} &1 - 2k\left(\frac{n}{k} - \frac{J_n'(k)}{J_n(k)}\right)^{-1} \\ &< 1 + \frac{2h\, J_0'(h)}{J_0(h)} - 4h\left(\frac{I_n'(h)}{I_n(h)} - \frac{J_n'(h)}{J_n(h)}\right)^{-1} , \qquad n = 2,3,4,\ldots \end{aligned}$$

For $n = 2$ the right hand side can be replaced by the smaller quantity -5; see 6.6 (c).

From 6.6 (20) we conclude that

$$(14) \qquad \frac{1}{4h}\left(\frac{I_n'(h)}{I_n(h)} - \frac{J_n'(h)}{J_n(h)}\right) > \sum_{\nu=1}^{\infty} j_{n\nu}^{-2} = \frac{1}{4(n+1)} .$$

Moreover, by 6.5 (11),

$$(15) \qquad \begin{aligned} \frac{1}{2k}\left(\frac{n}{k} - \frac{J_n'(k)}{J_n(k)}\right) &= \sum_{\nu=1}^{\infty} \frac{1}{j_{n\nu}^2 - k^2} \\ &< \sum_{\nu=1}^{\infty} j_{n\nu}^{-2} + 2k^2 \sum_{\nu=1}^{\infty} j_{n\nu}^{-4} \\ &= \frac{1}{4(n+1)} + \frac{2\,k^2}{16(n+1)^2\,(n+2)} \end{aligned}$$

provided $n \geqq 3$. Here we used the inequality

$$(1 - \alpha)^{-1} < 1 + 2\alpha, \qquad 0 < \alpha < 1/2,$$

and indeed

$$(k/j_{n\nu})^2 \leqq (k/j_{31})^2 = \left(\frac{3.8317}{6.3802}\right)^2 < 1/2 .$$

The following remains to be proved:

$$1 - \left(\frac{1}{4(n+1)} + \frac{2\,k^2}{16(n+1)^2(n+2)}\right)^{-1} < 6.27 - 4(n+1),$$

$$\left(\frac{1}{4(n+1)} + \frac{n+2}{2k^2}\right)^{-1} < 5.27 .$$

Replacing k by 4 and n by 3 this inequality holds indeed.

For n = 2 we use the second formula in (12) and obtain

$$1 - 2k\,\frac{J_2(k)}{J_3(k)} = -6.34 < -5 < -4.79 .$$

6.6 B. Biharmonic radii. We associate with a given curve certain radii which we call the biharmonic inner and biharmonic outer radius. They correspond to the ordinary inner and outer radii in the same way as the biharmonic equation corresponds to the Laplace or harmonic equation. The biharmonic inner radius s_a depends, besides the given curve, on a point a in the interior of this curve; the biharmonic outer radius $\bar{s}$ depends only on the given curve.

(a) Let C be the given curve, a an interior point and D the domain inside of C. The Green function $\Gamma(p)$ of the domain D with singularity at a is defined as follows: I. The function $\Gamma(p)$ has in the neighborhood of a the form $r^2 \log r + h(p)$ where r is the distance of the variable point p from a and h(p) is regular and biharmonic throughout D, i.e., it satisfies the biharmonic equation $\nabla^4 h = 0$. II. We have $\Gamma = 0$ and $\partial\Gamma/\partial n = 0$ on the curve C. Now the function h assumes a positive value* at a which we denote by $s_a^2/2$; then the positive quantity s_a is the biharmonic inner radius.

In the case of a circle of radius R we have

---

* This follows from a result of Hadamard (7, p. 28) according to which h(a) is an increasing set function, i.e., it increases when the curve C is replaced by another curve containing C in its interior.

(1) $$\Gamma(p) = |z - a|^2 \log\left|\frac{R(z - a)}{R^2 - \bar{a} z}\right| - \frac{1}{2R^2}\left(|R(z - a)|^2 - |R^2 - \bar{a}z|^2\right)$$

where z and a are the complex numbers representing the points p and a, and z = 0 is the center of the circle; hence

(2) $$s_a = \frac{R^2 - |a|^2}{R} .$$

(b) Further we define the biharmonic outer radius by considering the Green function $\bar{\Gamma}(p)$ of the domain $\bar{D}$ exterior to C. This function is defined as follows: I. In the neighborhood of the point at infinity it has the form

(3) $$\bar{\Gamma}(p) = \log\frac{1}{r} + a\, r^2 + bx + cy + \bar{h}$$

where $\bar{h}$ is finite and biharmonic ($z = x + iy$, $|z| = r$). II. On the curve C we have $\bar{\Gamma} = \partial\bar{\Gamma}/\partial n = 0$. Here a is a positive constant, b and c are real. The biharmonic outer radius $\bar{s}$ is defined by $a = (2\bar{s}^2)^{-1}$.

In the case of a circle of radius R we have

(4) $$\bar{\Gamma}(p) = \log\frac{R}{|z|} - \frac{R^2 - |z|^2}{2R^2} , \qquad \bar{s} = R .$$

6.6 C. The biharmonic radii of a nearly circular curve. Our purpose is to compute the outer and inner radii of a nearly circular curve $r = 1 + \rho(\varphi)$ (the notation is the same as throughout this chapter).* The inner radius s_c will be taken with respect to the centroid c; see 6.2 (1).

(a) We seek the Green function $\bar{\Gamma}$ of the exterior domain $r \geq 1 + \rho$ in the form

(1) $$\bar{\Gamma} = \log\frac{1}{r} - \frac{1}{2}(1 - r^2) + p(r, \varphi) + q(r, \varphi) ,$$

(2)
$$p(r, \varphi) = \sum_{n=0}^{\infty} r^{2-n}(A_n \cos n\varphi + B_n \sin n\varphi) + \sum_{n=0}^{\infty} r^{-n}(C_n \cos n\varphi + D_n \sin n\varphi),$$
$$q(r, \varphi) = \sum_{n=0}^{\infty} r^{2-n}(A_n' \cos n\varphi + B_n' \sin n\varphi) + \sum_{n=0}^{\infty} r^{-n}(C_n' \cos n\varphi + D_n' \sin n\varphi).$$

---

* For a more complete treatment see Hadamard 7.

The coefficients A_n, B_n, C_n, D_n of $p(r, \varphi)$ are of the first order, the coefficients A_n', B_n', C_n', D_n' of $q(r, \varphi)$ of the second order. We note that r^2u is biharmonic if u is a harmonic function. The outer radius $\bar{s}$ is obtained from

$$(2\bar{s}^2)^{-1} = 1/2 + A_0 + A_0' ,$$

(3) $$\bar{s} = 1 - A_0 - A_0' + 3 A_0^2/2 .$$

We write the boundary conditions in the form ($r = 1 + \rho$)

(4) $$\bar{\Gamma} = \rho^2 + p(1, \varphi) + \rho p_r(1, \varphi) + q(1, \varphi) = 0 ,$$

(5) $$\frac{\partial \bar{\Gamma}}{\partial r} = 2\rho - \rho^2 + p_r(1, \varphi) + \rho p_{rr}(1, \varphi) + q_r(1, \varphi) = 0 .$$

The first order terms yield

(6) $$p(1, \varphi) = 0 ,$$
$$2\rho + p_r(1, \varphi) = 0 ,$$

so that

(7) $$p(r,\varphi) = (1 - r^2)a_0 + 2 \sum_{n=1}^{\infty} (1 - r^2)r^{-n}(a_n \cos n\varphi + b_n \sin n\varphi) ;$$

in particular

(8) $$A_0 = -a_0 .$$

We take now the mean value of the second order terms in (4) and (5) and obtain

(9) $$\begin{cases} -a_0^2 - 2\sum_{n=1}^{\infty} (a_n^2 + b_n^2) + A_0' + C_0' = 0 , \\ -a_0^2 - 2 \sum_{n=1}^{\infty} (a_n^2 + b_n^2) -2a_0^2 + 2 \sum_{n=1}^{\infty} (4n - 2) (a_n^2 + b_n^2) + 2A_0' = 0 . \end{cases}$$

Consequently

(10) $$\begin{cases} A_0' = 3a_0^2/2 - \sum_{n=1}^{\infty} (4n - 3) (a_n^2 + b_n^2) , \\ C_0' = -a_0^2/2 + \sum_{n=1}^{\infty} (4n - 1) (a_n^2 + b_n^2) . \end{cases}$$

Using (3), (8) and the first equation in (10) we obtain for the biharmonic outer radius the following expression:

$$\bar{s} = 1 + a_0 + \sum_{n=1}^{\infty} (4n - 3)(a_n^2 + b_n^2) . \tag{11}$$

(b) We seek the Green function Γ_c of the domain $r \leqq 1 + \rho$ with respect to the centroid c in the following form:

$$\Gamma_c = r'^2 \log r' - \tfrac{1}{2}(r'^2 - 1) + p(r, \varphi) + q(r, \varphi), \tag{12}$$

$$\begin{cases} p(r,\varphi) = \sum_{n=0}^{\infty} r^{n+2}(A_n \cos n\varphi + B_n \sin n\varphi) + \sum_{n=0}^{\infty} r^n(C_n \cos n\varphi + D_n \sin n\varphi), \\ q(r,\varphi) = \sum_{n=0}^{\infty} r^{n+2}(A'_n \cos n\varphi + B'_n \sin n\varphi) + \sum_{n=0}^{\infty} r^n(C'_n \cos n\varphi + D'_n \sin n\varphi). \end{cases} \tag{13}$$

Here r' has the same meaning as in sect. 6.3. The coefficients of p are of the first order, those of q of the second order. The inner radius s_c is determined by

$$\begin{aligned} s_c^2/2 &= 1/2 + p(|c|, \gamma) + q(|c|, \gamma) \\ &= 1/2 + C_0 + (C_1 \cos \gamma + D_1 \sin \gamma)\,|c| + C'_0 , \end{aligned}$$

$$s_c = 1 + C_0 + (C_1 \cos \gamma + D_1 \sin \gamma)\,|c| + C'_0 - C_0^2/2 . \tag{14}$$

On the boundary $r = 1 + \rho$ we have

$$\begin{cases} r'^2 \log r' - \tfrac{1}{2}(r'^2 - 1) = [\rho - |c| \cos(\varphi - \gamma)]^2 , \\ 2r' \log r' \dfrac{dr'}{dr} = 2[\rho - |c| \cos(\varphi - \gamma)] + \rho^2 + |c|^2 - 2\rho|c| \cos(\varphi - \gamma), \end{cases} \tag{15}$$

so that

$$\Gamma_c = [\rho - |c| \cos(\varphi - \gamma)]^2 + p(1,\varphi) + \rho p_r(1,\varphi) + q(1,\varphi) = 0 , \tag{16}$$

$$\begin{aligned} \frac{\partial \Gamma_c}{\partial r} = 2[\rho - |c| \cos(\varphi - \gamma)] + \rho^2 + |c|^2 - 2\rho|c| \cos(\varphi - \gamma) \\ + p_r(1,\varphi) + \rho p_{rr}(1,\varphi) + q_r(1,\varphi) = 0 . \end{aligned} \tag{17}$$

The first order terms yield

$$(18)\quad \begin{cases} p(1,\varphi) = 0 , \\ 2[\rho - |c| \cos(\varphi - \gamma)] + p_r(1,\varphi) = 0 , \end{cases}$$

so that, in view of 6.2 (1),

$$(19)\quad p(r,\varphi) = (1 - r^2)\, a_0 + 2 \sum_{n=2}^{\infty} (1 - r^2)\, r^n (a_n \cos n\varphi + b_n \sin n\varphi) ;$$

in particular

$$(20)\qquad C_0 = a_0 , \qquad C_1 = D_1 = 0 .$$

Forming now the mean values of the second order terms we find

$$(21)\quad \begin{cases} a_0^2 + 2 \sum\limits_{n=2}^{\infty} (a_n^2 + b_n^2) - 2a_0^2 - 4 \sum\limits_{n=2}^{\infty} (a_n^2 + b_n^2) + A_0' + C_0' = 0 , \\ a_0^2 + 2 \sum\limits_{n=1}^{\infty} (a_n^2 + b_n^2) + |c|^2 - 4(a_1^2 + b_1^2) \\ -2a_0^2 - 4 \sum\limits_{n=2}^{\infty} (2n + 1)(a_n^2 + b_n^2) + 2 A_0' = 0 \end{cases}$$

so that

$$(22)\quad \begin{cases} A_0' = a_0^2/2 - a_1^2 - b_1^2 + \sum\limits_{n=2}^{\infty} (4n + 1)(a_n^2 + b_n^2) , \\ C_0' = a_0^2/2 + a_1^2 + b_1^2 - \sum\limits_{n=2}^{\infty} (4n - 1)(a_n^2 + b_n^2) . \end{cases}$$

Taking (14), (20) and the second equation in (22) into account we find

$$(23)\qquad s_c = 1 + a_0 + a_1^2 + b_1^2 - \sum_{n=2}^{\infty} (4n - 1)(a_n^2 + b_n^2) .$$

We note for comparison that the biharmonic inner radius s_0 of the nearly circular domain with respect to the origin is given by the expansion

$$(24)\qquad s_0 = 1 + a_0 - \sum_{n=1}^{\infty} (4n - 1)(a_n^2 + b_n^2) .$$

6.6 D. Comparison with other radii. We try to determine the relation of the biharmonic radii $\bar{s}$ and $\bar{s}_c$ to the quantities appearing in the table of 1.33.

Obviously (except for terms higher than the second order)

(1) $$\bar{s} \geqq \bar{r}$$

holds. On the other hand, $\bar{s} \leqq \bar{L}$ does not hold for an arbitrary variation since $n^2 \geqq 4n - 3$ is violated for $n = 2$. (It is correct for $n \geqq 3$.)

As to s_c we have the inequality (except for terms higher than the second order)

(2) $$s_c \leqq \bar{\Lambda}_2 .$$

On the other hand, $s_c \geqq \bar{\Lambda}_3$ does not hold for an arbitrary variation. We shall prove that the inequalities

(3) $$1 - 2k\left(\frac{n}{k} - \frac{J_n'(k)}{J_n(k)}\right)^{-1} < -(4n - 1) < 1 + \frac{2h\,J_0'(h)}{J_0(h)} - 4h\left(\frac{I_n'(h)}{I_n(h)} - \frac{J_n'(h)}{J_n(h)}\right)^{-1}$$

are satisfied for all $n \geqq 2$, excepting that the first inequality is reversed for $n = 2$.

(a) According to 6.6 (18) and 6.6 A (14), the extreme right expression in (3) is greater than

$$6.27 - 4(n + 1)$$

which is indeed greater than $-(4n - 1)$.

In view of 6.6 A (15), the other inequality in (3) can be written as follows:

$$\sum_{\nu=1}^{\infty} \frac{1}{j_{n\nu}^2 - k^2} < \frac{1}{4n} .$$

Using the inequality established in 6.6 A (15) it suffices to show that

$$\frac{1}{4(n + 1)} + \frac{2k^2}{16(n + 1)^2 (n + 2)} < \frac{1}{4n} ,$$

$$k < \left(\frac{2(n + 1)(n + 2)}{n}\right)^{1/2}$$

which is correct for $n \geqq 4$.

As to the cases $n = 2$ and $n = 3$, it is more convenient to use the second expression in 6.6 A (12). For $n = 2$ this is (as we have seen at the end of that sect.) $-6.34 > -7$. On the other hand

$$1 - 2k\frac{J_3(k)}{J_4(k)} = -11.62 < -11 \quad .$$

This proves the previous assertion concerning the inequalities (3).]

6.7. Tangential coordinates. (a) We follow the notation of 1.31, denoting by p, θ the tangential coordinates, so that the equation of the curve appears in the form $p = 1 + h(\theta)$. The transformation of the previous formulas written in polar coordinates r, φ into tangential coordinates p, θ can be carried out easily provided that we restrict ourselves to terms of not higher than the second order. Assuming that the poles and axes are the same, we find

$$(1) \quad \begin{cases} \varphi = \theta + h'(\theta) + \ldots \\ \rho(\varphi) = h(\theta) + \frac{1}{2}[h'(\theta)]^2 + \ldots \quad . \end{cases}$$

Consequently

$$(2) \quad a_0 = \frac{1}{2\pi}\int_{-\pi}^{\pi}\rho(\varphi)d\varphi = \frac{1}{2\pi}\int_{-\pi}^{\pi}(h(\theta) + \tfrac{1}{2}[h'(\theta)]^2)(1 + h''(\theta))d\theta$$

$$= \frac{1}{2\pi}\int_{-\pi}^{\pi}h(\theta)\,d\theta + \frac{1}{2\pi}\int_{-\pi}^{\pi}(hh'' + \tfrac{1}{2}(h')^2)d\theta$$

$$= \frac{1}{2\pi}\int_{-\pi}^{\pi}h(\theta)d\theta - \frac{1}{4\pi}\int_{-\pi}^{\pi}[h'(\theta)]^2d\theta$$

$$= h_0 - \sum_{n=1}^{\infty} n^2(h_n^2 + k_n^2) \ ;$$

we used 1.34 (2) and neglected terms of higher than the second order.

As to the quantities $a_n^2 + b_n^2$ they can be replaced by $h_n^2 + k_n^2$ without affecting the terms of the second order. This yields easily the general rule 1.34 (3).

(b) We can establish formulas analogous to (1) in space. We use the symbols θ, φ, $d\omega$ and θ', φ', $d\omega'$ in the same sense as in 1.34. In order to simplify matters we observe that our transformation affects only the mean values X_0 and Y_0. A relation between these mean values can easily be obtained by comparing, for instance, Hurwitz's formula* for V in tangential coordinates

* Sur quelques applications géométriques des séries de Fourier, Annales de l'École Normale Supérieure, series 3, vol. 19 (1902) pp. 357-408; Mathematische Werke, vol. 1, p. 554.

with the trivial formula

(3) $$V = \frac{1}{3} \iint r^3 \, d\omega = \frac{1}{3} \iint (1 + 3\rho + 3\rho^2) \, d\omega$$

valid in polar coordinates. This yields the relation

$$Y_0 - \sum_{n=2}^{\infty} \frac{(n+2)(n-1)}{8\pi} \iint Y_n^2 \, d\omega'$$

$$= X_0 + \sum_{n=1}^{\infty} \frac{1}{4\pi} \iint X_n^2 \, d\omega = X_0 + \sum_{n=1}^{\infty} \frac{1}{4\pi} \iint Y_n^2 \, d\omega'$$

so that

(4) $$X_0 = Y_0 - \sum_{n=1}^{\infty} \frac{n^2 + n}{8\pi} \iint Y_n^2 \, d\omega' .$$

This leads to the general rule 1.34 (7) which we shall use in deriving some of the results listed in the second part of the table 1.33.

6.8. Capacity. We seek Green's function G of the exterior domain $r > 1 + \rho$ in the form

(1) $$G = 1/r + p(r, \theta, \varphi) + q(r, \theta, \varphi)$$

(2) $$p(r,\theta,\varphi) = \sum_{n=0}^{\infty} S_n(\theta,\varphi) r^{-n-1}, \qquad q(r,\theta,\varphi) = \sum_{n=0}^{\infty} S_n'(\theta,\varphi) r^{-n-1} ,$$

where S_n and S_n' are surface harmonics of degree n with first and second order coefficients, respectively. The capacity is given by

(3) $$C = 1 + S_0 + S_0' .$$

We have $G = 1$ on $r = 1 + \rho$ so that

(4) $$1 = (1 + \rho)^{-1} + p(1 + \rho, \theta, \varphi) + q(1 + \rho, \theta, \varphi)$$

$$= 1 - \rho + \rho^2 + p(1,\theta,\varphi) + \rho \, p_r(1,\theta,\varphi) + q(1,\theta,\varphi)$$

$$= 1 - \rho + \rho^2 + \sum_{n=0}^{\infty} S_n - \rho \sum_{n=0}^{\infty} (n+1) S_n + \sum_{n=0}^{\infty} S_n' .$$

The terms of the first order yield $-\rho(\theta,\varphi) + p(1,\theta,\varphi) = 0$, and hence we obtain, using the notation of 1.33 (9), that

(5) $$p(r,\theta,\varphi) = \sum_{n=0}^{\infty} X_n(\theta,\varphi)\, r^{-n-1} .$$

We note in particular

(6) $$S_0 = X_0 .$$

Taking the mean value of the second order terms in (4), we find

(7) $$\sum_{n=0}^{\infty} \frac{1}{4\pi} \iint X_n^2\, d\omega - \sum_{n=0}^{\infty} \frac{n+1}{4\pi} \iint X_n^2\, d\omega + S_0' = 0 .$$

Now (3), (6) and (7) yield formula 1.33 (11), or no. 14 of the table 1.33.

6.9. Inner radius with respect to the centroid. Let (x,y,z) denote a variable point in space at distance r from the origin and put

$$x = r\xi , \qquad y = r\eta , \qquad z = r\zeta .$$

The first coordinate of the centroid c of a uniform volume distribution can be obtained as follows:

(1) $$\frac{3}{4\pi} \iiint r\xi \; r^2\, dr\, d\omega = \frac{3}{4\pi}\, \frac{1}{4} \iint (1+\rho)^4\, \xi\, d\omega$$
$$= \frac{3}{4\pi} \iint \rho\xi \; d\omega = A ,$$

provided that $X_1 = A\xi + B\eta + C\zeta$; here terms of higher than the first order have been neglected. Let r' be the distance of a variable point from c; we seek Green's function G_c with respect to c in the form

(2) $$G_c = 1/r' - 1 - p(r,\theta,\varphi) - q(r,\theta,\varphi)$$

where the coefficients of the harmonic functions

(3) $$p(r,\theta,\varphi) = \sum_{n=0}^{\infty} S_n(\theta,\varphi) r^n, \qquad q(r,\theta,\varphi) = \sum_{n=0}^{\infty} S_n'(\theta,\varphi) r^n$$

are of the first and second order, respectively. According to its definition (32), we find the inner radius r_c from the relation

(4) $$1/r_c = 1 + S_0 + (r\,S_1) + S_0'$$

where the term $(r\,S_1)$ has to be taken at c.

Let r_0 be the distance of the centroid c from the origin and γ the variable angle between the radii r and r_0. In view of (1) we have $r_0 \cos\gamma = X_1$. Now $G_c = 0$ on the boundary so that

$$(5)\quad 1/r' = (r^2 + r_0^2 - 2r_0\, r \cos \gamma)^{-1/2}$$

$$= \frac{1}{r} + \frac{r_0}{r^2} \cos \gamma + \frac{r_0^2}{r^3} P_2 (\cos \gamma) + \ldots$$

$$= 1 - \rho + \rho^2 + r_0(1 - 2\rho) \cos \gamma + r_0^2 P_2 (\cos \gamma) + \ldots$$

$$= 1 + p(1, \theta, \varphi) + \rho\, p_r(1, \theta, \varphi) + q(1, \theta, \varphi) ,$$

where P_n denotes Legendre's polynomial. We compare the terms of the first order:

$$(6)\qquad - \rho + r_0 \cos \gamma = p(1, \theta, \varphi) .$$

We conclude hence that

$$(7)\qquad S_0 = - X_0 , \qquad S_1 = 0$$

and

$$(8)\qquad p(r, \theta, \varphi) = - X_0 - \sum_{n=2}^{\infty} X_n(\theta, \varphi)\, r^n .$$

Taking the mean value of the terms of the second order in (5) and using (8), we find

$$(9)\quad \sum_{n=0}^{\infty} \frac{1}{4\pi} \iint X_n^2\, d\omega - 2 \frac{1}{4\pi} \iint X_1^2\, d\omega = - \sum_{n=2}^{\infty} \frac{n}{4\pi} \iint X_n^2\, d\omega + S_0' .$$

It follows from (4), (7) and (9) that

$$(10)\quad 1/r_c = 1 - X_0 + X_0^2 - \frac{1}{4\pi} \iint X_1^2\, d\omega + \sum_{n=2}^{\infty} \frac{n+1}{4\pi} \iint X_n^2\, d\omega .$$

This establishes no. 16 of the table 1.33.

6.10. Volume, surface-area and Minkowski's constant. These quantities' have been dealt with by A. Hurwitz who used tangential coordinates (loc. cit.). From his results and the general rule 1.34 (7) for the transition from tangential to polar coordinates, nos. 10, 11 and 15 of the table 1.33 follow without difficulty.

6.11. Mean values of the radii of a solid. In this section we return to the mean values N_λ defined in 3.3. As we have seen, N_λ is an increasing function of λ and $N_3 = \overline{V} \leqq C$. We prove now that the exponent 3 cannot be replaced by a larger one; more precisely, we show that for $\lambda > 3$ a nearly spherical domain can be found such that $N_\lambda > C$.

Indeed

$$(1) \quad N_\lambda = \left\{\frac{1}{4\pi}\iint(1+\rho)^\lambda\, d\omega\right\}^{1/\lambda}$$

$$= \left\{1 + \frac{\lambda}{4\pi}\iint \rho\, d\omega + \frac{\lambda(\lambda-1)}{8\pi}\iint \rho^2\, d\omega\right\}^{1/\lambda}$$

$$= 1 + \frac{1}{4\pi}\iint \rho\, d\omega + \frac{\lambda-1}{2}\left\{\frac{1}{4\pi}\iint \rho^2\, d\omega - \left(\frac{1}{4\pi}\iint \rho\, d\omega\right)^2\right\}$$

$$= 1 + X_0 + \sum_{n=1}^{\infty} \frac{\lambda-1}{8\pi}\iint X_n^2\, d\omega \ .$$

In view of 1.33 (11) we have

$$(2) \qquad C - N_\lambda = \sum_{n=1}^{\infty} \frac{n - \frac{1}{2}(\lambda - 1)}{4\pi}\iint X_n^2\, d\omega \ .$$

If we choose $\rho(\theta, \varphi) = \delta\bar{\rho}(\theta, \varphi) = \delta \cos\theta$, the right hand side of (2) will be negative for $\lambda > 3$.

6.12. <u>Discussion of another inequality</u>. We have (1.13 (2), 1.13 (4))

$$(1) \qquad \bar{V} \leqq C \leqq \bar{M}\ , \qquad\qquad \bar{M}^{1/4}\, \bar{V}^{3/4} \leqq \bar{S} \leqq \bar{M}\ .$$

Thus we can raise the question whether an inequality of the form

$$(2) \qquad \bar{M}^\lambda\, \bar{V}^\mu \leqq C\ , \qquad\qquad \lambda > 0,\ \mu > 0,\ \lambda + \mu = 1$$

holds where λ and μ are fixed values. We show that this inequality is false for <u>any</u> λ and μ satisfying the conditions mentioned, provided that a suitable nearly spherical domain is chosen.

We use nos. 10, 14 and 15 of the table 1.33 assuming $X_0 = 1$. Then

$$(3) \quad \log C - \lambda \log \bar{M} - \mu \log \bar{V} = \sum_{n=1}^{\infty} \frac{(n-1)(2\mu - \lambda n)}{8\pi}\iint X_n^2\, d\omega \ .$$

Choosing $\rho(\theta, \varphi) = \delta P_n(\cos\theta)$ where $n > 2\mu/\lambda$ and $n > 1$, and P_n is Legendre's polynomial, we render quantity (3) negative for sufficiently small δ.

In drawing the final conclusion here and in 6.11 we follow the same line of argument as in 1.35.

CHAPTER VII. ON SYMMETRIZATION*

7.1. Another definition of symmetrization. The apparently scattered remarks of the present chapter are, in fact, carefully grouped around the idea of symmetrization. We begin by connecting this idea with two general concepts concerning real functions of n real variables $x_1, x_2, \ldots, x_n$.**

We say that the functions $f(x_1,x_2,\ldots,x_n)$ and $g(x_1,x_2,\ldots,x_n)$ are similarly ordered if they are defined in the same n-dimensional domain R and if

$$[f(x_1,\ldots,x_n) - f(x_1',\ldots,x_n')][g(x_1,\ldots,x_n) - g(x_1',\ldots,x_n')] \geqq 0$$

for any two points $(x_1,x_2,\ldots,x_n)$ and $(x_1',x_2',\ldots,x_n')$ of R. We say that f and g are oppositely ordered if f and -g are similarly ordered.

We consider the set of those points at which

$$f(x_1,x_2,\ldots,x_n) \geqq u$$

and let M(u) denote the measure of this set. Let N(u) be so related to g as M(u) is to f. We say that the functions f and g are equimeasurable if for all real values of u

$$M(u) = N(u) .$$

In visualizing the meaning of these terms, we may think of sufficiently "regular" functions of two or three variables. If f(x,y) and g(x,y) are equimeasurable, the volumes under the surfaces representing these functions in the usual way are "equisectional", that is, their cross-sections with an arbitrary horizontal plane have the same area although not necessarily the

---

* [Some of the subjects discussed in 7.1 and 7.3 are also treated, more intuitively and less generally, in note A which may, but need not, be read first.]

** Compare G. H. Hardy, J. E. Littlewood and G. Pólya, Inequalities, 1934, theorem 236 and section 10.12.

same form. If $f(x,y,z)$ and $g(x,y,z)$ are similarly ordered, they have the same level surfaces and their gradients have the same direction although not necessarily the same magnitude.

We may use the terms just introduced to define three kinds of symmetrization: with respect to a plane (Steiner), with respect to a straight line (Schwarz), and with respect to a point. We consider functions of a variable point (x,y,z) which are defined in the whole space, take only real, non-negative values and vanish outside a certain bounded domain. With a given function $f = f(x,y,z)$ of this kind we associate three functions f', f'' and f''' which also depend on x, y and z. For any given x and y, f and f' are functions of the one variable z and are as such equimeasurable and f' and z^2 are oppositely ordered. For a given z, f and f'' are functions of the variables x and y and are as such equimeasurable, and f'' and $x^2 + y^2$ are oppositely ordered. Finally, f and f''' are equimeasurable as functions of all three variables x, y and z, and f''' and $x^2 + y^2 + z^2$ are oppositely ordered. These conditions imply that f', f'' and f''' depend in a special way on x, y and z:

$$f' = g_1(x,y,z^2)\ ,\quad f'' = g_2(x^2 + y^2,\ z)\ ,\quad f''' = g_3(x^2 + y^2 + z^2)\ .$$

The level surfaces of f' are symmetric with respect to the x,y-plane, those of f'' are surfaces of revolution about the z-axis and those of f''' are spheres with center at the origin. We say that f is transformed into f', f'' and f''' by three different kinds of symmetrization, with respect to the x,y-plane, the z-axis, and the origin, respectively.

With a given bounded set S in space we associate its characteristic function f which takes the value 1 at points of S and the value 0 at points outside S. The three kinds of symmetrization change S into S', S'' and S''' of which f', f'' and f''' are the characteristic functions, respectively. The first of these definitions is equivalent to that given in 1.7. The function f^* considered in 3.2 is essentially the same as f'''. The parallelism of the three kinds of symmetrization was emphasized in 24. We connect the transformation of S into S' and S'' with the names of Steiner and Schwarz, respectively.

Of course, we can also define symmetrization in a plane by using the terms "oppositely ordered" and "equimeasurable".

7.2. On moments of inertia. The definitions given in the foregoing section enable us to use the following simple, but general theorem:*

* See Hardy, Littlewood and Pólya, loc. cit., theorem 378. The proof given there applies to any number of variables.

If f, f^+, f^-, g, g^+ *and* g^- *are defined in the same bounded n-dimensional region,* f^+ *and* g^+ *are similarly ordered,* f^- *and* g^- *oppositely ordered,* f, f^+ *and* f^- *are equimeasurable, and also* g, g^+ *and* g^- *are equimeasurable, then*

$$(1) \quad \int \cdots \int f^- g^- dx_1 \ldots dx_n \leq \int \cdots \int f\, g\, dx_1 \ldots dx_n \leq \int \cdots \int f^+ g^+ dx_1 \ldots dx_n \,.$$

Each integration is extended over the common domain of definition.

We give just one simple application. Let D, I_0, I denote, as usual, a plane domain, its polar moment of inertia about the origin and about its center of gravity, respectively. Symmetrization with respect to the x-axis transforms D into D^*. Let D^*, I_0^*, and I^* be in the same relation to each other as D, I_0, and I. Let f and f^* be the characteristic functions of D and D^*, respectively. Then, for each fixed x, as functions of y alone, $f(x,y)$ and $f^*(x,y)$ are equimeasurable and $x^2 + y^2$ and $f^*(x,y)$ oppositely ordered. We suppose that the center of gravity of D is at the origin. Applying the first inequality of (1) when integrating with respect to y, we find that

$$\begin{aligned} I = I_0 &= \iint_D (x^2 + y^2) dx\, dy \\ &= \int \left[\int (x^2 + y^2) f\, dy \right] dx \\ &\geq \int \left[\int (x^2 + y^2) f^* dy \right] dx \\ &= \iint_{D^*} (x^2 + y^2) dx\, dy = I_0^* \geq I^* \,. \end{aligned}$$

We conclude that $I \geq I^*$. That is, *symmetrization diminishes the polar moment of inertia with respect to the center of gravity.* Cf. 20.

7.3. *A generalization of the theorem on symmetrization.* The most important part of the theorem stated in 1.10 is based on the argument presented in 24, pp. 6-8. This argument applies with little modification to a much more general situation. We wish, however, to confine ourselves (as in 24) to a relatively simple geometric configuration in order to emphasize the generality in other, perhaps less obvious, respects.

Let A_0 and A_1 denote closed surfaces; A_0 lies in the interior of A_1 and the domain between A_0 and A_1 is called the "field". We consider a function f which takes the value 0 on $\underline{A}_0$, the value 1 on $\underline{A}_1$ and values between 0 and 1 in the field. We assume that the function f behaves relatively simply. More precisely, we suppose: the points where $f = \lambda$ form, if $0 < \lambda < 1$, a closed surface $\underline{A}_\lambda$ (a level surface of f) contained in the interior of the field. If $0 \leq \alpha < \beta \leq 1$, $\underline{A}_\alpha$ is contained in the interior of $\underline{A}_\beta$. Through each interior point of the field passes just one well-defined surface $\underline{A}_\lambda$ with $0 < \lambda < 1$. In short, the level surfaces $\underline{A}_\lambda$ surround each other and fill the field completely; we suppose them also sufficiently smooth so as to admit the analytic operations that we shall perform later. (If f coincides within the field with the harmonic function that takes the prescribed values 0 and 1 on the surfaces $\underline{A}_0$ and $\underline{A}_1$, respectively, all our assumptions are legitimate.) We extend the definition of f to the whole space in setting:

$$f = 0 \quad \text{on and inside } \underline{A}_0, \quad f = 1 \quad \text{on and outside } \underline{A}_1 .$$

Symmetrization with respect to the x,y-plane, as defined in 7.1, transforms $1 - f$ and $\underline{A}_\lambda$ into $1 - f'$ and $\underline{A}'_\lambda$, respectively $(0 \leq \lambda \leq 1)$. Therefore, z^2 and f', considered as functions of z alone, are similarly ordered for any fixed x and y, and $f' = \lambda$ on $\underline{A}'_\lambda$ for $0 \leq \lambda \leq 1$. Also, $\underline{A}'_\alpha$ is contained in $\underline{A}'_\beta$ if $\alpha < \beta$.

(a) We consider a function F(t) defined for $t \geq 0$ and subject to the conditions:

(I) F(t) is convex (not concave) from below,

(II) F(t) is increasing (not decreasing).

Our aim is to <u>show under these assumptions that</u>

$$(1) \qquad \iiint F(|\text{grad } f|)dx\,dy\,dz \geq \iiint F(|\text{grad } f'|)dx\,dy\,dz ;$$

the first integral is extended over the field between $\underline{A}_0$ and $\underline{A}_1$, the second between $\underline{A}'_0$ and $\underline{A}'_1$.

We consider a (vertical) straight line perpendicular to the plane of symmetrization through the point (x,y,0). Let this line intersect the surface $\underline{A}_\lambda$ at the points

$$(x,y,z_1) , (x,y,z_2) , (x,y,z_3) , \ldots, (x,y,z_{2m})$$

where

$$z_1 > z_2 > z_3 > \ldots > z_{2m} .$$

The number $2m = 2m(x,y)$ of these points is finite. Let γ_μ denote the angle between the exterior normal to $\underline{A}_\lambda$ at the point (x,y,z_μ) and the positive z-axis. Then

$$\cos \gamma_\mu = (-1)^{\mu-1}\left\{1 + (\partial z_\mu/\partial x)^2 + (\partial z_\mu/\partial y)^2\right\}^{-1/2}$$

and, at the point considered,

$$|\text{grad } f| \cos \gamma_\mu = (\partial \lambda/\partial n)\cos \gamma_\mu = \partial \lambda/\partial z_\mu .$$

Therefore, the contribution of the neighborhoods of the 2m points considered to the left hand side of (1) is

$$(2) \qquad dx\, dy\, d\lambda \sum_{\mu=1}^{2m} F\left\{\frac{[1 + (\partial z_\mu/\partial x)^2 + (\partial z_\mu/\partial y)^2]^{1/2}}{|\partial z_\mu/\partial \lambda|}\right\}\left|\frac{\partial z_\mu}{\partial \lambda}\right| .$$

Observe that the vertical line considered intersects $\underline{A}'_\lambda$ at the points (x,y,z) and (x,y,-z) where

$$2z = z_1 - z_2 + z_3 - z_4 + \dots + z_{2m-1} - z_{2m} .$$

The neighborhoods of these two points contribute to the right hand side of (1) and their contribution corresponding to (2) is

$$(2') \qquad 2dx\, dy\, d\lambda\; F\left\{\frac{[1 + (\partial z/\partial x)^2 + (\partial z/\partial y)^2]^{1/2}}{|\partial z/\partial \lambda|}\right\}\left|\frac{\partial z}{\partial \lambda}\right| .$$

We introduce the symbols

$$(3) \qquad \begin{aligned} &|\partial z_\mu/\partial x| = p_\mu , \quad |\partial z_\mu/\partial y| = q_\mu , \quad (-1)^{\mu-1}\partial z_\mu/\partial \lambda = r_\mu , \\ &\sum_{\mu=1}^{2m} p_\mu = P , \quad \sum_{\mu=1}^{2m} q_\mu = Q , \quad \sum_{\mu=1}^{2m} r_\mu = R ; \end{aligned}$$

p_μ, q_μ and r_μ are positive. Rewriting (2) and (2') with this notation and with due regard for the expression of z, we find that the proof of (1) reduces to that of the following inequality:

$$(4) \qquad \sum_{\mu=1}^{2m} r_\mu F\left(\frac{(1 + p_\mu^2 + q_\mu^2)^{1/2}}{r_\mu}\right) \geq R\, F\left(\frac{(4 + P^2 + Q^2)^{1/2}}{R}\right) .$$

Now, as the function F(t) is convex,

$$(5)\qquad \sum_{\mu=1}^{2m} r_\mu F(a_\mu) \geq R\, F\left(R^{-1} \sum_{\mu=1}^{2m} r_\mu a_\mu\right) \quad ;$$

we set

$$(6)\qquad a_\mu = r_\mu^{-1}(1 + p_\mu^2 + q_\mu^2)^{1/2} \quad .$$

By a simple case of Minkowski's inequality

$$\sum_{\mu=1}^{2m} r_\mu a_\mu = \sum_{\mu=1}^{2m} (1 + p_\mu^2 + q_\mu^2)^{1/2}$$

$$(7)\qquad \geq \left\{\left(\sum_{\mu=1}^{2m} 1\right)^2 + \left(\sum_{\mu=1}^{2m} p_\mu\right)^2 + \left(\sum_{\mu=1}^{2m} q_\mu\right)^2 \right\}^{1/2}$$

$$\geq (4 + P^2 + Q^2)^{1/2} \quad .$$

As F(t) is also an increasing function of t, (5), (6) and (7) establish (4).

(b) We may observe that f and f' are equimeasurable in z for fixed x and y, and therefore also equimeasurable as functions of the three variables x, y and z. Hence, for an arbitrary function G(t),

$$(8)\qquad \iiint G(f)dx\,dy\,dz = \iiint G(f')dx\,dy\,dz$$

if the domain of integration is the interior of $\underline{A}_1$ on the left hand side and that of $\underline{A}_1'$ on the right hand side.

(c) Several quantities important in mathematical physics can be considered as the minimum of a quotient of the following kind:

$$(9)\qquad \frac{\iiint F(|\operatorname{grad} f|)dx\,dy\,dz}{H\{\iiint G(f)dx\,dy\,dz\}} \quad .$$

If $F(t) = t^2$, G(t) arbitrary, $H(t) = 4\pi$, and the geometrical configuration and the boundary values prescribed for f are as described, the capacity C of the condenser between the surfaces (coatings) A_0 and A_1 is the minimum of (9); compare 2.2 (8). If we pass from three to two dimensions, change suitably the configuration and the boundary values, and take first F(t), G(t) and H(t) as before, then

$$F(t) = t^2 , \quad G(t) = t , \quad H(t) = 4t^2 ,$$

and finally

$$F(t) = t^2 , \quad G(t) = t^2 , \quad H(t) = t ,$$

we obtain as the minimum c, 1/P and Λ^2, respectively; see sect. 2.3, 5.1 (2) and (3).

Steiner symmetrization decreases the numerator of (9) (see (1)), leaves the denominator unchanged (see (8)), and hence decreases also the quotient (9) or its minimum (see A 4 (a) for details). This remark proves immediately what the theorem of sect. 1.10 asserts concerning C, P, and Λ. The assertion concerning $\dot{r}$ and $\bar{r}$ follows, too, since these quantities are limiting cases of c. Concerning S and L see (1) with $F(t) = (1 + t^2)^{1/2}$, concerning V and A see (8) with G(t) = t. For I and M we must refer, however to sect. 7.2 and 7.9, respectively.

(d) Steiner symmetrization, repeated an infinity of times with respect to a suitably chosen sequence of planes intersecting in a given axis, generates Schwarz symmetrization with respect to that axis. [This indication will be somewhat expanded in sect. A 5.] By applying the theorem on Steiner symmetrization proved under (a) an infinity of times, we obtain in the limit an analogous theorem on Schwarz symmetrization. A direct proof has been given (24, pp. 8-11) for the most important particular case which yields the result that *Schwarz symmetrization diminishes the capacity*. A particular case of this theorem can be stated as follows: *The capacity of a cylindrical solid is never less than the capacity of a right circular cylinder that has the same volume and the same altitude as the given solid*. A limiting case of the last statement is the following: *Of all conducting plates with a given area the circle has the minimum electrostatic capacity*. This theorem, conjectured by Lord Rayleigh (26, vol. 2, p. 179) and first proved by the authors (24, p. 14), can also be expressed by the inequality

$$C \geq 2\pi^{-3/2} A^{1/2}$$

valid for the capacity C of a plate with area A.

7.4. *Simple cases*. Symmetrization, which means *Steiner symmetrization* in the present section, changes a polygon into a polygon, but usually increases the number of sides. The rather exceptional cases in which symmetrization does not increase the complexity of the figure are noteworthy; they lead to curious theorems some of which seem to be inaccessible to any other known method.

(a) Of all triangles with a given A, the equilateral triangle has the smallest L, I, $\bar{r}$, C and Λ, but the largest $\dot{r}$ and P.

All quantities named are continuous functions of the vertices of the triangle. It will suffice to show this for the outer radius $\bar{r}$. Let Δ be a triangle and k a positive number. Changing the sides of Δ in the ratio 1 : k, we change Δ into a similar triangle which we denote by $k\Delta$. Let $\bar{r}$ be the outer radius of Δ; then $k\bar{r}$ is the outer radius of $k\Delta$. Now let ε be an arbitrary positive number, $\varepsilon < 1$. If the vertices of a triangle Δ_1 are sufficiently near to those of Δ, the triangle Δ_1 can be enclosed between the triangles $(1 - \varepsilon)\Delta$ and $(1 + \varepsilon)\Delta$, and so its outer radius is contained between $(1 - \varepsilon)\bar{r}$ and $(1 + \varepsilon)\bar{r}$. This proves our assertion. We used, in fact, only the following very general properties of the outer radius and the triangle: $\bar{r}$ has a dimension (the dimension 1), $\bar{r}$ is a monotonic functional (its value for a subdomain cannot exceed its value for the full domain) and the triangle is star-shaped with respect to an (to any) inner point. Hence the argument for $\bar{r}$ applies, with small variations, also to L, I, $\dot{r}$, C, P and Λ.

We start now from an arbitrary triangle. Symmetrizing it with respect to one of its altitudes, we change it into an isosceles triangle. Symmetrizing this isosceles triangle with respect to the altitude perpendicular to one of its two equal sides, we change it into another isosceles triangle. Repeating this process, we attain ultimately, in the limit, after an infinite number of symmetrizations, an equilateral triangle with the given area A; this can be proved by an elementary discussion. Each of the quantities L, I, $\bar{r}$, $\dot{r}$, C, P and Λ changes monotonically during the process, see 1.10, and attains ultimately a limiting value, which is an extremum, as asserted.

The theorem just proved could also be expressed by seven inequalities of which we quote only two: for any triangle with area A, we have

$$\Lambda \geqq 2.3^{-1/4}\pi A^{-1/2}, \quad P \leqq 3^{1/2}A^2/15 .$$

(b) Of all quadrilaterals with a given A, the square has the smallest L, I, $\bar{r}$, C and Λ, but the largest $\dot{r}$ and P.

After the discussion in (a), it is sufficient to indicate a sequence of symmetrizations which transform, ultimately, a given quadrilateral into a square. Symmetrizing a given quadrilateral with respect to a perpendicular to one of its diagonals, we change it into a quadrilateral having a diagonal as axis of symmetry. Symmetrizing this new quadrilateral with respect to a perpendicular to its axis of symmetry, we change it into a rhombus. Symmetrizing the rhombus with respect to a perpendicular to one of its sides, we change it into a rectangle. Symmetrizing the rectangle with respect to a perpendicular to one of its diagonals, we obtain another rhombus. Repeating the last two steps in succession, we obtain an infinite sequence in which rhombi alternate with rectangles. The sequence converges to a square (which is both a rhombus and a rectangle) with the given area A. This proves the theorem and so, among others, the inequality

$$\Lambda \leqq \pi(2/A)^{1/2}$$

valid for the principal frequency of any quadrilateral with area A.

The proof based on symmetrization which works for the regular polygon with 3 or 4 sides (under (a) and the present (b), respectively) fails when we come to the regular pentagon. To prove (or disprove) the analogous theorems for regular polygons with more than four sides is a challenging task.

(c) _Of all tetrahedra with a given_ V, _the regular tetrahedron has the minimum_ S, M _and_ C.

We symmetrize the given tetrahedron with respect to a plane perpendicular to one of its edges. Then we symmetrize the tetrahedron just obtained with respect to a plane perpendicular to the edge that lies in the former plane of symmetrization; the tetrahedron so obtained has two planes of symmetry perpendicular to each other. By repeating these two steps appropriately, we attain ultimately a regular tetrahedron with the given volume V.

(d) _Of all prisms (right or oblique) with a given_ V _and a quadrilateral base, the cube has the minimum_ S, M _and_ C.

Symmetrizing with respect to a plane perpendicular to the lateral edges of the given prism, we change it into a right prism. We take now as successive planes of symmetrization planes perpendicular to the base of the right prism and passing through the lines of symmetrization used under (b). We arrive so finally at a right prism with square base. Now, interchange the roles: take a lateral face (a rectangle) as base and repeat the process.

(e) _Consider convex polyhedra with_ 8 _triangular faces and_ 6 _vertices in each of which four faces meet. Of all such polyhedra with a given_ V _the regular octahedron has the minimum_ S, M _and_ C.

Symmetrizing three times in succession with respect to planes, each of which is perpendicular to an appropriately chosen diagonal, we obtain a solid of the proposed kind with three mutually perpendicular planes of symmetry. We treat now the rhombus in which one of these planes intersects the solid as under (b), changing it finally into a square. Now, interchange the role of the planes of symmetry.

It seems difficult to prove analogous theorems for the (regular) dodecahedron or icosahedron.

(f) An ellipse with given A is steadily elongated. That is, the ellipse varies so that the ratio of its major axis to its minor axis steadily increases. Then L, I, $\bar{r}$, C and Λ increase, and $\dot{r}$ and P decrease monotonically.

We consider two ellipses with semi-axes a_1, b_1 and a_2, b_2, respectively, where

$$(1) \qquad a_1 b_1 = a_2 b_2 , \quad 1 \leqq a_1/b_1 < a_2/b_2 .$$

We symmetrize the ellipse (a_2, b_2) with respect to a variable line revolving about the center. The symmetrized figure is again an ellipse one axis of which lies on the line of symmetrization. The ratio of this axis to the other varies from a_2/b_2 to b_2/a_2 as the line of symmetrization turns through 90°. Therefore, this ratio must take somewhere the value 1 and so, by (1), also the value a_1/b_1. This means that, symmetrized with respect to a suitable line, the ellipse (a_2, b_2) goes over into the ellipse (a_1, b_1). This symmetrization changes L, I and the other quantities mentioned in a predictable direction (1.10) and so our assertion is proved.

See also sect. 1.9, 1.22 and B 5.

[(g) The following remarks, although not obtained by symmetrization, are in other respects similar to the foregoing. They are presented in the Master's Thesis of W. J. Perry (Stanford University, 1950). They are concerned with the volume V, the surface-area S, and the Minkowski constant M of prisms, pyramids, and double pyramids. (A double pyramid consists of two pyramids attached to opposite sides of the same base. If this base has n sides, the double pyramid has 2n faces.) Each of these three classes of polyhedra contains one regular polyhedron: the cube, the regular tetrahedron, and the regular octahedron, respectively. Let us say that a prism is "semiregular" if it is a right prism and its base is a regular polygon. A pyramid is semiregular if its base is a regular polygon and the foot of its altitude coincides with the center of its base. A double pyramid is semiregular if the two pyramids of which it consists are semiregular, each a mirror image of the other.

Of all prisms with a given number n of lateral faces and a given surface-area S, a certain semiregular prism has the maximum volume V and a certain semiregular prism has the minimum Minkowski constant M. These two prisms, however, are actually different, with one exception: for $n = 4$ they both coincide with the cube.

It is natural to expect similar results for pyramids with n lateral faces and double pyramids with 2n faces so that the exceptional coincidence arises for $n = 3$ and the regular tetrahedron among pyramids, and for $n = 4$ and the regular octahedron among double pyramids. This conjecture has been partly proved by W. J. Perry, namely insofar as it is concerned only with semiregular solids of both classes.]

7.5. On the inner radius of convex curves. (a) We prove the following theorem: If a convex curve has two axes of symmetry, the inner radius of the curve attains its maximum at the point of intersection of the axes.

Let c be the point of intersection of the two given axes of symmetry and a any other point in the interior of the curve. We assert that

$$r_a \leqq r_c . \tag{1}$$

For the proof we distinguish two cases.

If the two axes are perpendicular to each other (as in the case of a rectangle), we symmetrize first with respect to one of the axes, then with respect to the other. The curve remains unchanged both times, but a goes over into b, a point on the first axis, and then b into c. By the theorem on symmetrization (1.10)

$$r_a \leqq r_b \leqq r_c$$

which proves (1).

If the given axes of symmetry are not perpendicular (as in the case of an equilateral triangle), we symmetrize an infinity of times first with respect to the first axis, then with respect to the second, then again with respect to the first, and so on, alternately. The curve remains unchanged, but the point a goes over successively into b_1, b_2, b_3, b_4, ... The points b_1, b_3, b_5, ... lie on the first axis, b_2, b_4, b_6, ... on the second, and the distances of b_1, b_2, b_3, ... from c decrease in geometric progression so that

$$\lim_{n \to \infty} b_n = c . \tag{2}$$

By the theorem on symmetrization

$$r_a \leqq r_{b_1} \leqq r_{b_2} \leqq r_{b_3} \leqq \cdots . \tag{3}$$

Yet r_a is a continuous function of a, and so from (2) and (3) we infer (1).

(b) [The foregoing proof deserved to be mentioned, because it yields without essential change also the corresponding result in three dimensions. By conformal mapping, however, it is easy to prove the theorem announced in 1.23 which contains the result derived in (a). We consider the domain D in the complex z-plane and assume that D is convex and admits a rotation of

$2\pi/m$ about $z = 0$; $m \geq 2$. Then D is the conformal image of a circle $|\zeta| < r_o$ under a mapping of the form

(4) $$z = f(\zeta) = \zeta + c_{m+1}\zeta^{m+1} + c_{2m+1}\zeta^{2m+1} + \dots .$$

The radius r_o of this circle is the inner radius of D with respect to the origin. We prove that for any position of the point a in the interior of D

(5) $$r_a \leqq r_o \quad ;$$

<u>if equality is attained, we have either</u> $a = 0$, <u>or the domain D is an infinite strip between two parallel lines and</u> a (still a center with $m = 2$) <u>a point at equal distance from both parallels</u>.

Let $\zeta = \varphi(z)$ be the inverse function of $z = f(\zeta)$, and let the points $z = a$ and $\zeta = \alpha$ correspond to each other so that

$$\varphi(a) = \alpha \ , \ f(\alpha) = a \ .$$

An elementary argument shows* that

(6) $$r_a = \frac{r_o^2 - |\varphi(a)|^2}{r_o|\varphi'(a)|}$$

so that the inequality (5) can be written as follows:

(7) $$\frac{1}{|\varphi'(a)|} = |f'(\alpha)| \leqq \frac{1}{1 - |\alpha|^2 r_o^{-2}} \ .$$

We have to prove that the last inequality holds for all values of α, $|\alpha| < r_o$, $\alpha \neq 0$. We can assume that $r_o = 1$.

For the proof we note** that in view of the convexity of D the function

$$1 + \zeta \frac{f''(\zeta)}{f'(\zeta)}$$

is regular and has a positive real part for $|\zeta| < 1$. The constant term of this function being 1, the moduli of the higher coefficients are all $\leqq 2$.***

* Cf. G. Pólya-G. Szegö, Aufgaben und Lehrsätze, vol. 2, p. 19, problem 110.

** Loc. cit., vol. 1, p. 105, problem 108.

*** Loc. cit., vol. 1, p. 129, problem 235.

Since only terms of the form ζ^{km} occur we have

(8) $$1 + \zeta \frac{f''(\zeta)}{f'(\zeta)} \ll \frac{1 + \zeta^m}{1 - \zeta^m}$$

where the notation $\sum a_n \zeta^n \ll \sum A_n \zeta^n$ indicates that $|a_n| \leqq A_n$, $n = 0,1,2,\ldots$. We conclude

$$\frac{f''(\zeta)}{f'(\zeta)} \ll \frac{2\zeta^{m-1}}{1 - \zeta^m} ,$$

$$\log f'(\zeta) \ll -\frac{2}{m} \log(1 - \zeta^m) ,$$

$$f'(\zeta) \ll (1 - \zeta^m)^{-2/m} .$$

Using the inequality

$$(1 - \rho^m)^{-2/m} \leqq (1 - \rho^2)^{-1} , \quad 0 \leqq \rho < 1 , \quad m \geqq 2 ,$$

we finally find that

(9) $$|f'(\alpha)| \leqq (1 - |\alpha|^2)^{-1}$$

which is the desired (7), since $r_o = 1$. Equality holds, of course, if $\alpha = 0$. If equality holds for $\alpha \neq 0$, assuming $f'(0) > 0$, we find that $m = 2$,

$$f'(\zeta) = (1 - \zeta^2)^{-1}$$

and hence

$$z = f(\zeta) = \frac{1}{2} \log \frac{1 + \zeta}{1 - \zeta}$$

which function gives as image of the unit circle $|\zeta| < 1$ the strip

$$-\pi/4 < \Im z < \pi/4 .$$

A modification of the foregoing argument shows that inside any convex domain, symmetrical or not, there is just one point at which r_a is stationary; see 9.]

7.6. On the surface radius of convex solids. We introduce our problem by discussing first the analogous question in two dimensions.

(a) We consider a bounded and closed plane set $\underline{D}$ and its orthogonal projection onto an arbitrary straight line of its plane. There is a straight line ℓ for which this projection attains its maximum length p. We let $\bar{r}$ denote the transfinite diameter of $\underline{D}$. (This deviates from our usual notation; we extend now the use of $\underline{D}$ and $\bar{r}$ to more general sets; see 6.)

Without essential loss of generality, we assume that the boundary of $\underline{D}$ consists of a finite number of smooth curves. Symmetrizing $\underline{D}$ with respect to ℓ, we obtain a domain $\underline{D}'$ with transfinite diameter $\bar{r}'$. The line ℓ is an axis of symmetry for $\underline{D}'$ and intersects it along a finite number of segments the joint length of which is p. Symmetrizing $\underline{D}'$ with respect to a line perpendicular to ℓ, we obtain a domain $\underline{D}''$ with transfinite diameter $\bar{r}''$. The intersection of $\underline{D}''$ with the line ℓ is a segment of length p, the transfinite diameter of which is p/4, as is well known, and p/4 cannot be greater than $\bar{r}''$. By our general theorem on symmetrization

$$\bar{r} \geqq \bar{r}' \geqq \bar{r}'' \geqq p/4 \ . \tag{1}$$

We found thus a new proof for a known result; see Table 1.21, entry 32.*

(b) Assume now that $\underline{D}$ is convex and let $p(\theta)$ denote the length of the orthogonal projection of $\underline{D}$ onto a line that includes the angle θ with a fixed direction. Then

$$L = \int_0^{\pi} p(\theta)\, d\theta \leqq \pi p$$

and so we obtain from (1)

$$\bar{r} L^{-1} \geqq (4\pi)^{-1} \ . \tag{2}$$

A lower bound of $\bar{r}L^{-1}$ for convex curves can be obtained also by the inclusion lemma. The ratio $\bar{r}L^{-1}$ has obviously no positive lower bound for non-convex curves. There are reasons to suspect that, for convex curves, the right hand side of (2) can be replaced by 1/8; this bound would be attained in the limiting case of a line segment (the "needle").

(c) We consider now a bounded and closed set $\underline{B}$ in space and its orthogonal projection onto an arbitrary plane. There is a plane ε for which the area of this projection attains its maximum value $\underline{P}$. We let C denote the capacity of $\underline{B}$, as usual.

---

* G. Pólya, Beitrag zur Verallgemeinerung des Verzerrungssatzes auf mehrfach zusammenhängende Gebiete, Sitzungsberichte der Preussischen Akademie der Wissenschaften, 1928, pp. 228-232.

Without essential loss of generality we admit that the boundary of $\underline{B}$ consists of a finite number of smooth surfaces. Symmetrizing $\underline{B}$ with respect to ε, we obtain a solid $\underline{B}'$ with capacity C'. The plane ε is a plane of symmetry for $\underline{B}'$ and the area of the intersection of $\underline{B}'$ with ε is $\underline{P}$. Symmetrizing $\underline{B}'$ with respect to a straight line perpendicular to ε (see 7.1), we obtain a solid of revolution $\underline{B}''$ with capacity C''. The intersection of $\underline{B}''$ with the plane ε is a circular disk with area $\underline{P}$, the capacity of which is $(2/\pi)(\underline{P}/\pi)^{1/2}$; this quantity, of course, cannot be greater than C''. By our general theorem on symmetrization

(3) $$C \geq C' \geq C'' \geq (2/\pi)(\underline{P}/\pi)^{1/2} .$$

We have proved entry 23 of the Table 1.21.

(d) Assume now that $\underline{B}$ is convex. Let ζ denote a variable vector drawn from the center of the unit sphere to its surface, $d\omega$ an element of this surface, and $P(\zeta)$ the area of the orthogonal projection of $\underline{B}$ onto a plane of which the normal is ζ. Then

$$\pi S = \iint P(\zeta)\, d\omega \quad ;$$

the integral is over the full unit sphere. Hence we obtain

$$\pi S \leq 4\pi \underline{P}$$

and so from (3)

(4) $$C/\bar{S} \geq 2/\pi .$$

This inequality is remarkable, since the ratio $C/\bar{S}$ has obviously no positive lower bound for non-convex surfaces. There are reasons to suspect that, for convex surfaces, the right hand side of (4) can be replaced by $8^{1/2}/\pi$; this bound would be attained in the limiting case of the circular disk.

7.7. On the shape of equipotential surfaces. We consider the electrostatic field around a charged conductor $\underline{B}$. Let $u = u(x,y,z)$ denote the electrostatic potential; u satisfies Laplace's equation outside $\underline{B}$, takes the value 1 on the surface $\underline{A}$ of $\underline{B}$ and the value 0 at infinite distance. We consider the equipotential surface with the equation

(1) $$u(x,y,z) = \lambda$$

and call it $\underline{A}(\lambda)$. Observe that $0 < \lambda \leq 1$ and that $\underline{A}(1) = \underline{A}$. If $\lambda' > \lambda$, $\underline{A}(\lambda')$ is contained in the interior of $\underline{A}(\lambda)$.

We have the intuitive idea that the level surfaces "become steadily more spherical" as they expand. There are various possibilities of giving a precise meaning to this intuitive statement; we choose an interpretation that we are able to prove. Let $V(\lambda)$, $S(\lambda)$ and $C(\lambda)$ denote the volume, the surface-area, and the capacity of $\underline{A}(\lambda)$, respectively. If $\underline{A}(\lambda)$ were a sphere, the quantity

$$3V(\lambda) \; / \; 4\pi[C(\lambda)]^3 \tag{2}$$

would have the value 1. If $\underline{A}(\lambda)$ is not a sphere (in fact $\underline{A}(\lambda)$ cannot be a sphere unless $\underline{A}$ itself is one) the deviation of the value of (2) from 1 indicates the deviation of $\underline{A}(\lambda)$ from the spherical shape. We shall prove the following precise interpretation of the above intuitive statement: <u>As λ decreases and $\underline{A}(\lambda)$ expands, the ratio (2) increases steadily from a certain stage onward and tends to 1.</u>

That (2) tends to 1 as λ approaches 0, follows from a routine discussion of the expansion of u in spherical harmonics; we omit this part of the proof. The crucial part of the assertion is that <u>the derivative of the function (2) of λ is negative for sufficiently small λ.</u> We shall prove this assertion by changing appropriately Szegö's proof for Poincaré's theorem (table 1.21, entry 2); compare sect. 3.2.

We need to know here that $\underline{A}(\lambda)$ is a convex surface for sufficiently small λ. We only hint the proof. We consider the Gaussian curvature of the equipotential surface passing through the point (x,y,z) the distance of which from the origin is r. We expand this curvature in powers of 1/r (with homogeneous polynomials in x/r, y/r, z/r as coefficients) and find that the leading term is $1/r^2$. Hence follows the convexity of $\underline{A}(\lambda)$ in points sufficiently distant from the origin.

We observe here that, when $\underline{A}(\lambda)$ is convex, it makes sense to consider $M(\lambda)$, the integral of the mean curvature extended over the surface $\underline{A}(\lambda)$.

In accordance with 2.1 (2), the capacity of $\underline{A}$ is

$$C = \frac{1}{4\pi} \iint - \frac{\partial u}{\partial n} \, d\sigma \quad . \tag{3}$$

The integration is extended over an arbitrary closed surface containing $\underline{A}$, for instance, over $\underline{A}$ itself or over $\underline{A}(\lambda)$. From the last remark we easily see that

$$C(\lambda) = C/\lambda \quad . \tag{4}$$

Let dn denote the element of the common normal intercepted by the neighboring equipotential surfaces $\underline{A}(\lambda)$ and $\underline{A}(\lambda + d\lambda)$. Then

$$dV(\lambda) = \iint dn \, d\sigma = d\lambda \iint d\sigma \, / [\partial\lambda / \partial n]$$

and so

$$(5) \qquad -V'(\lambda) = -dV(\lambda)/d\lambda = \iint \frac{d\sigma}{-\partial u/\partial n} \quad ;$$

the integration is extended over $\underline{A}(\lambda)$. Observe that the integrand is positive both in (3) and in (5) (n is the exterior normal). Applying Schwarz's inequality to (3) and (5), we obtain

$$(6) \qquad -V'(\lambda) 4\pi C \geqq [S(\lambda)]^2 \quad .$$

In view of (4), the expression (2) is equal, except for a positive constant factor, to $V(\lambda)\lambda^3$. Therefore, we have to examine the derivative of $V(\lambda)\lambda^3$ for sufficiently small λ. We choose λ so small that $\underline{A}(\lambda)$ is convex and $M(\lambda)$ has a meaning. Using successively (6), the second of the inequalities 1.13 (4) of Minkowski, then (4) and finally the inequality 1.13 (2) of Szegö, provèd in 3.4, we obtain

$$\begin{aligned}
-[V(\lambda)\lambda^3] &= -V'(\lambda)\lambda^3 - 3V(\lambda)\lambda^2 \\
&\geqq \frac{[S(\lambda)]^2\lambda^3}{4\pi C} - 3V(\lambda)\lambda^2 \\
&\geqq \frac{3M(\lambda)V(\lambda)\lambda^3}{4\pi C} - 3V(\lambda)\lambda^2 \\
&= \frac{3V(\lambda)\lambda^3}{C}\left[\frac{M(\lambda)}{4\pi} - C(\lambda)\right] \\
&\geqq 0 \quad .
\end{aligned}$$

With this we have proved that the deviation of the equipotential surfaces from the spherical form, as indicated by (2), diminishes steadily as these surfaces expand beyond a certain stage.

7.8. On the second proper frequency of a vibrating membrane. The principal frequency of a membrane, that is, its first or lowest proper frequency, is diminished by symmetrization, and hence we may quite reasonably suspect that the higher frequencies behave similarly. Yet there is a less obvious heuristic reason to suspect the contrary. In fact, if symmetrization would consistently diminish the proper frequencies of all membranes with a given area, the circular membrane should have the minimum proper frequency of any given rank. Yet the proper frequencies or "eigenvalues" of the circle, except the lowest, are not simple. When, however, a double root is decomposed by slight variation into two simple roots, one of these is usually greater, the other less, than the original double root which, in such a case, is neither a maximum nor a minimum.*

The following simple example shows that this heuristic argument points in the right direction. We consider a rectangle with sides a, b, area A, and second proper frequency Λ_2. Then

$$A = ab \ , \ \Lambda_2^2 = \pi^2(a^{-2} + 4b^{-2})$$

provided that $a \leqq b$. If A is given, the minimum of Λ_2 is attained when $b = 2a$ and is equal to

$$\pi\ [a^{-2} + 4(2a)^{-2}]^{1/2} = \pi\, 2^{1/2}\, a^{-1} = 2\pi A^{-1/2} \ .$$

The second proper frequency of a circle with area A is

$$3.83\, \pi^{1/2}\, A^{-1/2} = 6.78\, A^{-1/2} > 2\pi A^{-1/2} \ .$$

Thus, for a given area, the second proper frequency of a circle is not a minimum. Hence we conclude that there are membranes the second proper frequency of which is increased by symmetrization.

7.9. On the integral of the mean curvature. Several examples led us to suspect that Steiner symmetrization diminishes M, the integral of the mean curvature extended over the surface of a convex solid. ("Diminishes" means also here "does not increase".) We proved this conjecture in some special cases. [And we communicated it to several mathematicians. The conjectured theorem has been proved then completely by H. Busemann, I. Fáry, and W. Fenchel, independently of each other.

We are given the convex solid $\underline{B}$. Steiner symmetrization with respect to a given plane p changes $\underline{B}$ into $\underline{B}^*$. The surface parallel to the surface of $\underline{B}$ at the distance h bounds the solid $\underline{B}_h$. Steiner symmetrization with respect to p transforms $\underline{B}_h$ into $(\underline{B}_h)^*$ and the surface parallel to that of $\underline{B}^*$ at the distance h bounds the solid $(\underline{B}^*)_h$.

Lemma. The solid $(\underline{B}^*)_h$ is contained in the solid $(\underline{B}_h)^*$.

In proving this lemma, we shall make use of the following obvious remarks:

---

* We are indebted to Dr. M. Schiffer for the foregoing remark.

If the solids $\underline{S}$ and $\underline{T}$ are transformed by symmetrization with respect to the same plane into $\underline{S}^*$ and $\underline{T}^*$, respectively, and $\underline{S}$ is contained in $\underline{T}$, also $\underline{S}^*$ is contained in $\underline{T}^*$.

If the solid $\underline{S}$ is convex and has a plane of symmetry parallel to the plane of symmetrization which changes it into $\underline{S}^*$, the solids $\underline{S}$ and $\underline{S}^*$ are congruent.

Let P be an arbitrary interior point of $\underline{B}$. The perpendicular to p through P has two points in common with the boundary of $\underline{B}$ which we call Q and R; the line-segment QR belongs to the convex solid $\underline{B}$. Let the center of a sphere with radius h describe the segment QR. Then the sphere itself describes a convex solid which we call $\underline{S}$ (the "sausage"); $\underline{S}$ consists of a right circular cylinder and of two hemispheres of radius h so connected that each base of the cylinder is the equatorial plane of a hemisphere; $\underline{S}$ has a plane of symmetry parallel to p. By definition, $\underline{S}$ is contained in $\underline{B}_h$. Symmetrization with respect to the given plane p changes P, Q, R and $\underline{S}$ into P^*, Q^*, R^* and $\underline{S}^*$, respectively. The segment Q^*R^*, congruent with QR, is contained in $\underline{B}^*$ by definition. The solid $\underline{S}^*$, congruent with $\underline{S}$, is contained in $(\underline{B}_h)^*$, by our initial remarks. Observe that the point P^*, in the same position on the segment Q^*R^* as P is on QR, can be <u>any</u> point of $\underline{B}^*$ and that the sphere with center P^* and radius h is contained in $\underline{S}^*$ (as the corresponding sphere in $\underline{S}$) and, therefore, in $(\underline{B}_h)^*$. Yet the whole solid $(\underline{B}^*)_h$ is made up of spheres with radius h centered in the points P^* of $\underline{B}^*$, and so the whole solid $(\underline{B}^*)_h$ is contained in $(\underline{B}_h)^*$, as asserted by the lemma.

As usual, let V, S and M denote the volume, the surface-area and the Minkowski constant of the solid $\underline{B}$, respectively, V_h the volume of $\underline{B}_h$ and let V^*, S^*, M^* and $(V^*)_h$ stand for the corresponding quantities connected with $\underline{B}^*$. Steiner symmetrization leaves the volume unchanged. Therefore

$$V^* = V \tag{1}$$

and V_h, the volume of $\underline{B}_h$, is also the volume of $(\underline{B}_h)^*$. Therefore, by the lemma,

$$(V^*)_h \leqq V_h \quad . \tag{2}$$

It is well known* that

$$V_h = V + Sh + Mh^2 + 4\pi h^3/3 \quad . \tag{3}$$

Similarly

$$(V^*)_h = V^* + S^*h + M^*h^2 + 4\pi h^3/3 \quad . \tag{4}$$

---

* Cf. W. Blaschke, Kreis und Kugel, 1916, p. 102.

From (1), (2), (3), and (4)

(5) $$S^*h + M^*h^2 \leqq Sh + Mh^2 \quad .$$

It follows hence, for $h \to 0$ and for $h \to \infty$,

(6) $$S^* \leqq S \ , \quad M^* \leqq M \quad .$$

The first inequality is a part of Steiner's well known result (we deal here only with convex solids), the second proves the conjecture stated at the outset.

It may be observed that, in opposition to other cases (see 3.2 and A 1) we proved only that M as a whole is diminished by symmetrization, but we did not prove (and it is not true) that the "elements" of M are so diminished.]

7.10. On spherical symmetrization. In order to point out a direction in which geometric operations similar to symmetrization may be sought, we state a conjecture: Of all domains on a given spherical surface with a given spherical area, the spherical bowl has the minimum electrostatic capacity.

[For a proof see note C.]

CHAPTER VIII. ON ELLIPSOID AND LENS

8.1. On the expansion of functions depending on the shape of an ellipsoid. The remarks of the present chapter are only loosely connected with each other. The first three sections deal with the ellipsoid, the last three with the lens; the results are broadly hinted in sections 1.36 and 1.37, respectively.

The present section discusses functions F(a,b,c) which depend only on the shape of the ellipsoid, not on its size; a, b and c denote the semi-axes. An example of such a function is the relative error of the approximation of C, the capacity of the ellipsoid, by $\bar{S}$, its surface-radius; that is, $(\bar{S}-C)/C$. Generally, such functions are characterized by three properties: the function F(a,b,c), defined for all positive values of the variables a, b and c, is

(I) symmetric,
(II) homogeneous of degree 0, and
(III) developable into a power series around the point (1,1,1).

If we suppose that the semi-axes a, b and c are decreasingly ordered, we can introduce α, β and γ, the numerical eccentricities of the three principal sections of the ellipsoid (defined by 1.36 (2) which supposes 1.36 (1)) and expand F(a,b,c) into powers of β and γ, on the basis of property (III). The initial term of the expansion of F(a,b,c), a function with the properties (I), (II) and (III), is never of second degree, and is necessarily equal to

$$[\beta^4],\ [\beta^6],\ [\beta^4]^2 \text{ or } [\beta^4][\beta^6]$$

multiplied by some constant factor, according as its degree is 4, 6, 8 or 10.

We have defined the "initial term" of an expansion and the abbreviations $[\beta^4]$ and $[\beta^6]$ in section 1.36; see 1.36 (4) and 1.36 (5). Our statement leaves aside the cases in which the degree of the initial term is 12 or higher; our method applies to any degree, but the result is less simple when the degree is greater than 10. In proving our statement, we shall restrict ourselves to the case of degree 4; this will be sufficient to indicate the method.

If $a > b$ and $a > c$, there are positive numbers β and γ satisfying two of the relations 1.36 (2). By (II)

$$(1) \qquad F(a,b,c) = F(1,\ b/a,\ c/a) = F(1,\ (1-\gamma^2)^{1/2},\ (1-\beta^2)^{1/2}) .$$

This can be expanded in powers of β^2 and γ^2, by (III), and the expansion must be symmetric in β and γ, by (I).

If $a > c$ and $b > c$, there are positive numbers α and β satisfying two of the relations 1.36 (2). By (II)

$$(2) \qquad \begin{aligned} F(a,b,c) &= F(c/b,\ c/a,\ (c/a)(c/b)) \\ &= F((1-\alpha^2)^{1/2},\ (1-\beta^2)^{1/2},\ (1-\beta^2)^{1/2}(1-\alpha^2)^{1/2}) . \end{aligned}$$

This can be expanded in powers of α^2 and β^2, by (III), and the expansion must be symmetric in α and β, by (I).

If $a > b > c$, both expansions are valid for sufficiently small α, β and γ. Moreover, by 1.36 (3),

$$(3) \qquad \gamma^2 = \beta^2 - \alpha^2 + (\beta^2 - \alpha^2)\alpha^2 + (\beta^2 - \alpha^2)\alpha^4 + \dots$$

and substituting this, we can pass from the first expansion, in powers of β^2 and γ^2, to the second, in powers of α^2 and β^2.

Assume now that the initial term of the first expansion is of degree 4. Since it is symmetric, it must be of the form

$$(4) \qquad p(\beta^4 + \gamma^4) + q\beta^2\gamma^2$$

where p and q are numerical coefficients. The substitution (3) allows us to pass from (4) to the initial term of the second expansion

$$(5) \qquad p\alpha^4 + (2p + q)\beta^4 - (2p + q)\alpha^2\beta^2$$

which must be also symmetric. Therefore

$$p = 2p + q ,$$

$q = -p$, and so (4) turns out to be of the form

$$p(\beta^4 - \beta^2\gamma^2 + \gamma^4) = p[\beta^4]$$

as asserted.

The rule which we have proved in the case of degree 4 (and the other cases are similar) was extremely useful in constructing the table in section 1.36. By the rule, the degree of the initial term determines this term, except for a constant factor, provided that the degree $\leqq 10$. Yet to find the degree and the constant factor, it is enough to consider just one kind of spheroid, oblate or prolate, and the corresponding expansions in powers of one eccentricity only. And so we can avoid altogether computing the more troublesome expansions in powers of two variables.

8.2. An approximation to the capacity of the ellipsoid. We consider entry 8 in the table of section 1.36. We shall prove that this approximation yields consistently too large values for prolate spheroids and too small values for oblate spheroids. The initial term of the relative error indicates this behavior but, of course, does not prove it.

(a) Prolate spheroids. In this case the inequality in question is [see 24, p. 22, (8.2)]

$$11\left(1+2(1-\beta^2)^{1/2}\right) + 4\left((1-\beta^2)^{1/2} + 2(1-\beta^2)^{1/4}\right) - 90\beta\left(\log\frac{1+\beta}{1-\beta}\right)^{-1}$$

(1)
$$= 11(3-\beta^2-\tfrac{1}{4}\beta^4-\tfrac{1}{8}\beta^6-\dots) + 4(3-\beta^2-\tfrac{5}{16}\beta^4-\tfrac{11}{64}\beta^6-\dots)$$
$$-45(1-\tfrac{1}{3}\beta^2-\tfrac{4}{45}\beta^4-\tfrac{44}{945}\beta^6-\dots)$$
$$= \frac{11}{336}\beta^6 + \dots > 0.$$

In order to prove this we write

(2) $$u = (1-\beta^2)^{1/4}, \quad \varphi(u) = 11 + 8u + 26u^2,$$

so that our inequality obtains the form:

(3)
$$1/2 \log\frac{1+\beta}{1-\beta} - \frac{45\beta}{\varphi(u)}$$
$$= \int_0^\beta \left(\frac{1}{1-\beta^2} - \frac{45}{\varphi(u)} + \frac{45\beta\varphi'(u)}{[\varphi(u)]^2}\frac{du}{d\beta}\right) d\beta > 0.$$

We prove the stronger statement that the integrand (which is a power series in β beginning with β^6) is positive, that is,

$$(4)\qquad \frac{1}{u^4} - \frac{45}{\varphi(u)} - \frac{45}{2}\,\frac{1-u^4}{u^3}\,\frac{\varphi'(u)}{(\varphi(u))^2} > 0\,,$$

or

$$(5)\qquad [\varphi(u)]^2 - 45u^4\,\varphi(u) - \frac{45}{2}\,u(1-u^4)\,\varphi'(u) > 0\,, \quad 0 < u < 1\,.$$

The last expression is a polynomial of degree 5 in u having a zero of order 6 at $\beta = 0$ or, which is the same, a zero of order 3 at $u = 1$. Hence it has the form

$$(6)\qquad (1-u)^3(A_0 + A_1 u + A_2 u^2)\,.$$

Comparing the terms 1, u, u^2 we obtain

$$(7)\qquad A_0 = 121,\quad A_1 = 359,\quad A_2 = 180$$

which renders the assertion obvious.

(b) Oblate spheroids. The inequality is in this case:

$$(8)\qquad \begin{aligned} &11(2 + (1-\beta^2)^{1/2}) + 4(2(1-\beta^2)^{1/4} + 1) - 45\beta(\arcsin\beta)^{-1} \\ &= 11(3 - \tfrac{1}{2}\beta^2 - \tfrac{1}{8}\beta^4 - \tfrac{1}{16}\beta^6 - \ldots) + 4(3 - \tfrac{1}{2}\beta^2 - \tfrac{3}{16}\beta^4 - \tfrac{7}{64}\beta^6 - \ldots) \\ &\qquad - 45(1 - \tfrac{1}{6}\beta^2 - \tfrac{17}{360}\beta^4 - \tfrac{367}{15120}\beta^6 - \ldots) \\ &= -\tfrac{11}{336}\beta^6 - \ldots < 0\,. \end{aligned}$$

Writing

$$(9)\qquad v = (1-\beta^2)^{1/4}, \quad \psi(v) = 26 + 8v + 11v^2\,,$$

we have to prove

$$(10)\qquad \arcsin\beta - \frac{45\beta}{\psi(v)} < 0$$

or the stronger inequality

$$(11)\qquad \begin{aligned} &(1-\beta^2)^{-1/2} - \frac{45}{\psi(v)} + \frac{45\beta\,\psi'(v)}{[\psi(v)]^2}\,\frac{dv}{d\beta} \\ &= v^{-2} - 45[\psi(v)]^{-1} - \frac{45}{2}\,v^{-3}(1-v^4)\,\psi'(v)[\psi(v)]^{-2} < 0\,. \end{aligned}$$

We write this inequality in the form

$$(12) \qquad v[\Psi(v)]^2 - 45v^3\Psi(v) - \frac{45}{2}(1 - v^4)\Psi'(v) < 0.$$

This is a polynomial of degree 5 in v having a zero of order 3 at $v = 1$. Hence it has the form

$$(13) \qquad (1 - v)^3 \; (B_0 + B_1 v + B_2 v^2) \quad .$$

Comparing the two lowest terms, with 1 and v, we find $B_0 = -180$, $B_1 = -359$. Comparing the highest terms, with v^5, we see that also $B_2 < 0$. This proves the assertion.

Of course, the inequalities (5) and (12) are essentially equivalent. Indeed, writing $u = v^{-1}$, we have in view of (2) and (9),

$$\varphi(u) = v^{-2}\psi(v) \quad .$$

Now multiplying (5) by v^5 we obtain (12). After the same operation, (6) yields

$$(v - 1)^3(A_0 v^2 + A_1 v + A_2) = (1 - v)^3(B_0 + B_1 v + B_2 v^2) \quad .$$

8.3. <u>Another approximation to the capacity of the ellipsoid</u>. We consider now entry 4 of the table in section 1.36 and prove a result analogous to that of the foregoing section: This approximation has also an error of constant sign for each kind of spheroids, prolate and oblate, but these two signs are opposite.

(a) <u>Prolate spheroids</u>. The following relation holds between the quantities M and S of an ellipsoid [24, p. 22, (7.24)]:

$$(1) \qquad M(a,b,c) = S((ab/c)^{1/2}, (ca/b)^{1/2}, (bc/a)^{1/2}).$$

Hence in the prolate case, when $b = c$, we have:

$$(2) \qquad M = M(a,b,b) = S(a^{1/2}, a^{1/2}, a^{-1/2}b) \quad .$$

The last quantity is the surface-area of an oblate spheroid with semi-axes $a' = b' = a^{1/2}$, $c' = a^{-1/2}b$. The corresponding numerical eccentricity is

$$(3) \qquad \beta' = (1 - c'^2/a'^2)^{1/2} = (1 - c^2/a^2)^{1/2} = \beta \ ,$$

so that

$$(4) \qquad M = \pi a \left(2 + (1 - \beta^2)\beta^{-1} \log \frac{1 + \beta}{1 - \beta} \right) \quad .$$

Moreover

(5) $$\bar{V} = (ab^2)^{1/3} = a(1-\beta^2)^{1/3}\ , \quad \bar{M} = M/(4\pi)\ ,$$

so that the inequality in question will be:

$$a^{-1}[\tfrac{1}{2}(\bar{V}+\bar{M}) - C]$$

$$= \tfrac{1}{2}(1-\beta^2)^{1/3} + \tfrac{1}{8}\left[2 + (1-\beta^2)\beta^{-1}\log\tfrac{1+\beta}{1-\beta}\right] - 2\beta\left(\log\tfrac{1+\beta}{1-\beta}\right)^{-1}$$

(6) $$= \tfrac{1}{2}(1-\tfrac{1}{3}\beta^2-\tfrac{1}{9}\beta^4-\tfrac{5}{81}\beta^6-\dots) + \tfrac{1}{8}(4-\tfrac{4}{3}\beta^2-\tfrac{4}{15}\beta^4-\tfrac{4}{35}\beta^6-\dots)$$

$$-(1-\tfrac{1}{3}\beta^2-\tfrac{4}{45}\beta^4-\tfrac{44}{945}\beta^6-\dots)$$

$$= \frac{4}{2835}\beta^6 + \dots > 0\ .$$

The left-hand expression of this inequality is decreasing if the logarithmic expression is decreased. But we have proved in 8.2 (a) that

(7) $$\log\frac{1+\beta}{1-\beta} > \frac{90\beta}{N}\ , \quad N = 11 + 8(1-\beta^2)^{1/4} + 26(1-\beta^2)^{1/2}\ ,$$

so that it suffices to prove the following inequality:

(8) $$\tfrac{1}{2}(1-\beta^2)^{1/3} + \tfrac{1}{4} + \tfrac{45}{4}(1-\beta^2)/N - N/45 > 0\ .$$

Writing $1-\beta^2 = u^{12}$, we obtain

(9) $$\frac{u^4}{2} + \frac{1}{4} + \frac{45}{4}\,\frac{u^{12}}{11+8u^3+26u^6} - \frac{11+8u^3+26u^6}{45} > 0\ .$$

Clearing the fractions, we obtain the following polynomial of degree 12 having a triple zero at $u = 1$:

$$-679u^{12} + 2340u^{10} - 1664u^9 + 720u^7 - 1374u^6 + 990u^4 - 344u^3 + 11$$

(10) $$= (1-u)^3(11 + 33u + 66u^2 - 234u^3 + 123u^4 + 1137u^5 + 1434u^6$$

$$+ 1734u^7 + 2037u^8 + 679u^9)\ .$$

We have to prove its positivity for $0 < u < 1$; for this purpose we prove that the polynomial of degree 9 in the parentheses is positive for all $u > 0$. We distinguish two cases:

1) $u < 0.6$; then

$$234\ (0.6)^3 < 11 + 33\ (0.6) + 66\ (0.6)^2 \ ;$$

2) $u > 0.6$; then

$$234 < 1137\ (0.6)^2$$

This proves in fact that the polynomial (10) is positive for $0 < u < 1$ and negative for $u > 1$.

(b) Oblate spheroids. In this case we have $a = b$ and

$$M = M(a,a,c) = S(ac^{-1/2},\ c^{1/2},\ c^{1/2}) \tag{11}$$

which is the surface-area of a prolate spheroid so that we obtain as above:

$$M = M(a,a,c) = 2\pi a[(1-\beta^2)^{1/2} + \beta^{-1}\arcsin\beta] \ . \tag{12}$$

Moreover

$$\bar{V} = (a^2c)^{1/3} = a(1-\beta^2)^{1/6}\ ,\quad \bar{M} = M/(4\pi)\ . \tag{13}$$

Thus the inequality in question will be

$$\begin{aligned} & a^{-1}[\tfrac{1}{2}(\bar{V}+\bar{M}) - C] \\ &= \tfrac{1}{2}(1-\beta^2)^{1/6} + \tfrac{1}{4}[(1-\beta^2)^{1/2} + \beta^{-1}\arcsin\beta] - \beta(\arcsin\beta)^{-1} \\ &= \tfrac{1}{2}(1-\tfrac{1}{6}\beta^2 - \tfrac{5}{72}\beta^4 - \tfrac{55}{1296}\beta^6 - \dots) + \tfrac{1}{4}(2-\tfrac{1}{3}\beta^2 - \tfrac{1}{20}\beta^4 - \tfrac{1}{56}\beta^6 - \dots) \\ &\qquad -(1-\tfrac{1}{6}\beta^2 - \tfrac{17}{360}\beta^4 - \tfrac{367}{15120}\beta^6 - \dots) \\ &= -\tfrac{4}{2835}\beta^6 - \dots < 0 \ . \end{aligned} \tag{14}$$

The left-hand expression is increasing if $\arcsin\beta$ is increased. But we have proved in 8.2 (b) that

$$\arcsin\beta < \frac{45\beta}{K}\ ,\qquad K = 26 + 8(1-\beta^2)^{1/4} + 11(1-\beta^2)^{1/2}\ , \tag{15}$$

so that it suffices to prove the following inequality:

$$\tfrac{1}{2}(1-\beta^2)^{1/6} + \tfrac{1}{4}[(1-\beta^2)^{1/2} + 45/K] - K/45 < 0 \ . \tag{16}$$

Writing $1-\beta^2 = u^{12}$, we obtain

$$(17). \qquad \frac{1}{2}u^2 + \frac{u^6}{4} + \frac{45}{4}\frac{1}{26 + 8u^3 + 11u^6} - \frac{26 + 8u^3 + 11u^6}{45} < 0 \quad .$$

Replacing here u by $1/u$ and multiplying by u^6, we regain the inequality (9) with the opposite sign. This was proved above.

8.4. On the capacity of a lens. The case of two contacting spheres.

Concerning the capacity C of a lens in general and in various special cases, cf. 32. We follow the notation of 1.37.

(a) The lens is a surface of revolution. Let $\bar{r}$ denote the outer radius of the meridian section which is a domain bounded by two circular arcs. We consider the interesting ratio

$$(1) \qquad C/\bar{r} .$$

As a consequence of the general result proved in 3.7 we can state that this quantity can never exceed $4/\pi$. After we have pointed out this result to Mr. J. G. Herriot (in particular in the case of a convex lens) he obtained (10) the much sharper result that for all lenses

$$(2) \qquad \frac{4}{\pi}\log 2 \leqq C/\bar{r} \leqq \frac{4}{\pi} \quad ;$$

the extreme values correspond to two equal tangent spheres and the circular disk, respectively.

In the next section we shall study in a direct manner the ratio (1) in the case of two mutually exclusive spheres tangent to each other. We prove in this case the inequalities

$$(3) \qquad \frac{4}{\pi}\log 2 \leqq C/\bar{r} \leqq 1$$

where equality is attained for two equal tangent spheres and for a single sphere, respectively.

(b) For general lenses the following conjecture can be formulated. Let $\bar{S}$ denote as usual the surface-radius. Then

$$(4) \qquad C \leqq \bar{S} .$$

For the spherical bowl we can verify this easily. Indeed, denoting by r the radius and by 2φ the central angle of the circular arc which is the meridian section of the bowl, we have (32, p. 348)

$$C = \pi^{-1} r\,(\varphi + \sin\varphi) \quad ,$$

$$S = 4\pi\bar{S}^2 = 4\pi r^2\,(1 - \cos\varphi) \quad ,$$

so that

$$C/\bar{S} \leqq 2^{-1/2}.4/\pi < 1 \quad .$$

Again we prove inequality (4) in case of two spheres tangent to each other, equality holding for the sphere.

Mr. Herriot obtained also in this respect more (10). He determined for the case of two tangent spheres not only the maximum but also the minimum of the ratio $C/\bar{S}$, showing that

$$(5) \qquad 2^{1/2} \log 2 \leqq C/\bar{S} \leqq 1$$

holds; these bounds are attained in the same cases as in (3). It is remarkable that $2^{1/2} \log 2 = 0.980$ so that in the case of two tangent spheres $C/\bar{S}$ is equal to 1 with an error less than 0.02. Finally, he proved (4) for a symmetrical lens and also in the case where $\alpha + \beta = \pi/2$.

8.5. Capacity and outer radius for two spheres in contact. Let us denote the radii of the spheres tangent to each other by r and r'. Then

$$(1) \qquad \bar{r} = \frac{\pi r r'}{(r + r') \sin \frac{\pi r}{r + r'}} \; ,$$

$$(2) \qquad C = \frac{rr'}{r + r'} \left\{ - \frac{\Gamma'}{\Gamma}\left(\frac{r}{r + r'}\right) - \frac{\Gamma'}{\Gamma}\left(\frac{r'}{r + r'}\right) - 2\gamma \right\}$$

where γ is Euler's constant [32, p. 346, (4)]. Hence

$$(3) \qquad \frac{C}{\bar{r}} = \frac{\sin \pi x}{\pi} \left\{ - \frac{\Gamma'}{\Gamma}(x) - \frac{\Gamma'}{\Gamma}(1-x) - 2\gamma \right\}, \quad x = \frac{r}{r + r'} \quad .$$

Since

$$- \frac{\Gamma'}{\Gamma}(x) - \gamma = \frac{1}{x} + \sum_{n=1}^{\infty} \left(\frac{1}{n + x} - \frac{1}{n}\right)$$

we have

$$(4) \qquad \frac{C}{\bar{r}} = \frac{\sin \pi x}{\pi} \left\{ \frac{1}{x} + \frac{1}{1 - x} + \sum_{n=1}^{\infty} \left(\frac{1}{n + x} + \frac{1}{n + 1 - x} - \frac{2}{n}\right) \right\}$$

$$= \frac{\sin \pi x}{\pi y} \left\{ 1 - \sum_{n=1}^{\infty} \frac{y(n + 2y)}{n(y + n + n^2)} \right\} \quad .$$

where $x(1 - x) = y$. Now*

$$\frac{\sin \pi x}{\pi y} = \prod_{n=1}^{\infty} \left(1 + \frac{y}{n(n+1)}\right). \tag{5}$$

Writing $[n(n + 1)]^{-1} = a_n$, we obtain

$$C/\bar{r} = \prod_{n=1}^{\infty} (1+a_n y) - \sum_{n=1}^{\infty} (1+2n^{-1} y) a_n y \prod_{\nu=1}^{n-1} (1+a_\nu y) \prod_{\nu=n+1}^{\infty} (1+a_\nu y) \tag{6}$$

$$= 1 - \sum_{n=1}^{\infty} a_n y \prod_{\nu=1}^{n-1} (1+a_\nu y) \left\{-1 + (1+2n^{-1} y) \prod_{\nu=n+1}^{\infty} (1+a_\nu y)\right\}$$

$$= 1 - \sum_{n=2}^{\infty} b_n y^n$$

where $b_n > 0$; we have used a well known transformation of an infinite product into a series. Consequently, the ratio $C/\bar{r}$ is a decreasing function of y. This furnishes the inequalities 8.4 (3).

---

* This remarkable identity and the obvious conclusion that

$$\sin \pi x = \sum_{n=1}^{\infty} A_n y^n$$

has positive coefficients A_n, has been pointed out by I. Schur. Cf. G. Pólya - G. Szegö, Aufgaben und Lehrsätze, vol. 1, p. 128, problems 227, 228. The conclusion mentioned follows also from the following formula of Lommel [cf. Watson, Bessel functions, p. 140, formula 5.22, (3)]:

$$(\pi z/2)^{-1/2} \cos(z^2 - 2zt)^{1/2} = \sum_{n=0}^{\infty} \frac{t^n}{n!} J_{n-1/2}(z)$$

by writing $z = \pi/2$, $t = \pi y$, so that

$$A_n = \frac{1}{2} \frac{\pi^{n+1}}{n!} J_{n-1/2}(\pi/2)$$

We have to use the well-known representation

$$J_\alpha(z) = \frac{(z/2)^\alpha}{\Gamma(\alpha+1/2)\Gamma(1/2)} \int_{-1}^{1} (1-t^2)^{\alpha-1/2} \cos zt \, dt, \quad \alpha > -1/2 .$$

8.6. Capacity and surface radius for two spheres in contact. In order to prove the inequality 8.4 (4) for two spheres of radii r, r' tangent to each other, we point out that

$$\bar{S} = (r^2 + r'^2)^{1/2} = (r + r')(1 - 2y)^{1/2} \tag{1}$$

so that following the notation of the previous section,

$$C/\bar{S} = (1 - 2y)^{-1/2} \left\{ 1 - \sum_{n=1}^{\infty} \frac{y(n + 2y)}{n(y + n + n^2)} \right\} . \tag{2}$$

We have to prove the inequality

$$\sum_{n=1}^{\infty} \frac{(2y + n)y}{n(y + n + n^2)} > 1 - (1 - 2y)^{1/2} = y + \sum_{n=2}^{\infty} h_n y^n$$

where $h_n > 0$. Another form of this inequality is:

$$y^{-1} \sum_{n=1}^{\infty} \left\{ \frac{2y+n}{n(y+n+n^2)} - \frac{1}{n(n+1)} \right\} = \sum_{n=1}^{\infty} \frac{2n+1}{n(n+1)(y+n+n^2)} > \sum_{n=2}^{\infty} h_n y^{n-2} .$$

The left-hand side is decreasing whereas the right-hand side is increasing. Hence it suffices to prove it in the special case $y = 1/4$, $x = 1/2$ which is the case of two equal spheres, $r = r'$. In this case

$$C = 2r \log 2, \quad \bar{S} = 2^{1/2} r, \quad C/\bar{S} = 2^{1/2} \log 2 < 1 .$$

This establishes the assertion.

NOTE A. SURFACE-AREA AND DIRICHLET'S INTEGRAL

A 1. Steiner symmetrization of the surface-area. The definition of Steiner symmetrization and a comprehensive theorem on it have been formulated in sect. 1.7-1.10. The present note gives an intuitive proof for a substantial part of the theorem of sect. 1.10. This proof is independent of the more abstract and more general developments of ch. 7 (especially 7.1 and 7.3) and yields also certain related results.

We shall make essential use of the following facts, discovered by Steiner:

I. Steiner symmetrization leaves the volume unchanged.

II. Steiner symmetrization diminishes (does not increase) the surface area.

We begin by proving these statements. We consider a given solid $\underline{B}$ and its surface $\underline{A}$ in a system of rectangular coordinates x, y, z. We take the (vertical) y,z-plane as plane of symmetrization. Let a (horizontal) straight line parallel to the x-axis intersect the surface $\underline{A}$ at the points

$$(x_1, y, z), \quad (x_2, y, z), \quad \ldots \quad (x_{2m}, y, z)$$

where

$$x_1 > x_2 > \ldots > x_{2m} \quad ,$$

$m \geq 1$. Steiner symmetrization with respect to the y, z-plane changes $\underline{A}$ and $\underline{B}$ into $\underline{A}^*$ and $\underline{B}^*$, respectively. By definition (sect. 1.7), to the intersection of $\underline{B}$ with the straight line just considered there corresponds in $\underline{B}^*$ a line-segment bisected by the y,z-plane, the endpoints of which have the coordinates x,y,z and -x,y,z, respectively, where

$$(1) \qquad 2x = x_1 - x_2 + x_3 - x_4 + \ldots + x_{2m-1} - x_{2m} \quad .$$

We express now the volume V of $\underline{B}$:

$$(2) \qquad V = \iint (x_1 - x_2 + x_3 - x_4 + \ldots + x_{2m-1} - x_{2m}) dy\, dz \quad .$$

The domain of integration is the orthogonal projection of B onto the plane of symmetrization. Of course, $m = m(y,z)$ depends on the choice of the point (y,z) in this plane. The points in which $m(y,z)$ is not defined may be neglected on the right hand side of (2) if the surface A is sufficiently smooth. (We may assume A as consisting of a finite number of analytic pieces or even as a polyhedral surface and use some approximation argument afterwards.)

The solid B^* is bisected by the plane of symmetrization and so its volume is

$$V^* = 2 \iint x \, dy \, dz \quad . \tag{3}$$

The domain of integration is the same as for (2) and x is given by (1). Therefore,

$$V^* = V \tag{4}$$

and statement I is proved.

We express now the surface-area S of A:

$$S = \iint \sum_{\nu=1}^{2m} \left[1 + \left(\frac{\partial x_\nu}{\partial y} \right)^2 + \left(\frac{\partial x_\nu}{\partial z} \right)^2 \right]^{1/2} dy \, dz \quad . \tag{5}$$

The domain of integration is the same as for (2) and the same comments on $m = m(y,z)$ apply here. The surface A^* is bisected by the plane of symmetrization and its area S^* can be expressed as an integral with the same domain of integration as in (5):

$$\begin{aligned} S^* &= 2 \iint \left[1 + \left(\frac{\partial x}{\partial y} \right)^2 + \left(\frac{\partial x}{\partial z} \right)^2 \right]^{1/2} dy \, dz \\ &= \iint \left[4 + \left(\sum_{\nu=1}^{2m} (-1)^{\nu-1} \frac{\partial x_\nu}{\partial y} \right)^2 + \left(\sum_{\nu=1}^{2m} (-1)^{\nu-1} \frac{\partial x_\nu}{\partial z} \right)^2 \right]^{1/2} dy \, dz \ ; \end{aligned} \tag{6}$$

we used (1).

Consider the vector with the components

$$1 \ , \ (-1)^{\nu-1} \frac{\partial x_\nu}{\partial y} \ , \ (-1)^{\nu-1} \frac{\partial x_\nu}{\partial z}$$

for $\nu = 1,2,\ldots 2m$. The sum of these 2m vectors has as first component $2m \geq 2$. As the length of the sum of these vectors cannot exceed the sum of the lengths, the integrand in (6) cannot exceed the integrand in (5), and so

$$S^* \leq S \quad . \tag{7}$$

This establishes statement II.

We can observe here a parallelism between the proofs of (4) and (7). We showed not only that the volume as a whole is preserved by symmetrization, but even that each element of it (each element of the integral (2)) is preserved. Similarly, not only the surface-area as a whole is diminished by symmetrization, but each element of it is diminished (transition from the integrand in (5) to that in (6)). In studying the effects of symmetrization on integrals of various kinds, we can often observe a similar behavior; compare sect. 3.2.

We shall need one more fact.

III. Steiner symmetrization leaves unchanged the area of the intersection with any plane perpendicular to the plane of symmetrization.

In fact, such an area (and, more generally, the area of the intersection of $\underline{B}$ with any cylinder whose generators are parallel to the x-axis) is represented, before and after symmetrization, by an integral of the form

$$\int (x_1 - x_2 + x_3 - x_4 + \dots + x_{2m-1} - x_{2m})ds$$

where ds denotes a line element in the plane of symmetrization.

A 2. Steiner symmetrization of a function. We consider now a domain $\underline{D}$ in the (horizontal) x,y-plane and a function f(x,y) which is positive in the interior of $\underline{D}$ and vanishes along the boundary of $\underline{D}$. We consider the points (x,y,z) where (x,y) belongs to $\underline{D}$ and

$$0 \leq z \leq f(x,y) \quad .$$

The set of such points is a solid $\underline{B}$. The boundary of $\underline{B}$ consists of two parts: its "base" which is the domain $\underline{D}$ in the x,y-plane, and its "upper surface" represented by the equation

$$(1) \qquad z = f(x,y) \quad .$$

We consider also the intersection of $\underline{B}$ with the (horizontal) plane z = t, where $t \geq 0$, and call A(t) the area of this intersection; A(0) is the area of $\underline{D}$, A(t) is a decreasing function of t, and A(t) = 0 for $t > t_1$, if t_1 denotes the maximum of f(x,y).

Steiner symmetrization with respect to the y,z-plane changes $\underline{B}$ into $\underline{B}^*$, the base $\underline{D}$ of $\underline{B}$ into the base $\underline{D}^*$ of $\underline{B}^*$, and the upper surface of $\underline{B}$ into the upper surface of $\underline{B}^*$. We write the equation of the upper surface of $\underline{B}^*$ in the form

$$(2) \qquad z = f^*(x,y) \quad ;$$

here we used the fact that the boundary of B and, therefore, also that of $\underline{B}^*$ has at most two points in common with any line parallel to the z-axis.

The function $f^*(x,y)$ is positive in the interior of $\underline{D}^*$ and vanishes along the boundary of $\underline{D}^*$; considered as a function of x alone (for fixed y) it is an even function and not increasing when $|x|$ increases. We shall say that $f(x,y)$ is changed into $f^*(x,y)$ by Steiner symmetrization with respect to the y,z-plane.

By the last remark of sect. A 1, the area $A(t)$ remains unchanged: the area of the intersection with the (horizontal) plane $z = t$ is the same for both surfaces, (1) and (2). Let $G(t)$ be any function continuous for $t \geq 0$; we assert that

$$\iint_{\underline{D}} G[f(x,y)]dx\,dy = \iint_{\underline{D}^*} G[f^*(x,y)]dx\,dy \quad . \tag{3}$$

In fact, both sides of (3) are obviously equal to

$$-\int_0^{t_1} G(t)dA(t) \quad .$$

Steiner symmetrization changes $G[f(x,y)]$ into $G[f^*(x,y)]$ if it changes $f(x,y)$ into $f^*(x,y)$. (We may assume that $G(0) = 0$ and $G(t) > 0$ as $t > 0$, or consider a slightly more general meaning of the term "symmetrization of a function".) Therefore, (3) can be considered as contained in theorem I of sect. A 1.

A 3. Transition from the surface-area to Dirichlet's integral. We apply now the result of sect. A 1 concerning the surface-area to the case considered in sect. A 2. The area of the whole surface of $\underline{B}$ is diminished by symmetrization. Yet the area of the base $\underline{D}$ remains unchanged (equal to $A(0)$), as we have observed (sect. A 1, III) and so the area of the upper surface must diminish:

$$\iint_{\underline{D}} [1 + f_x^{\,2} + f_y^{\,2}]^{1/2}dx\,dy \geq \iint_{\underline{D}^*} [1 + f_x^{*2} + f_y^{*2}]^{1/2}dx\,dy \quad . \tag{1}$$

The subscripts denote partial derivatives, $f_x = \partial f/\partial x$, etc. (The expression of the surface-area differs from that given in A 1(5).) Let us apply inequality (1) to εf, instead of f, where ε is an arbitrary positive constant. We obtain

$$\iint_{\underline{D}} [1 + \varepsilon^2(f_x^2 + f_y^2)]^{1/2}dx\,dy \geq \iint_{\underline{D}^*} [1 + \varepsilon^2(f_x^{*2} + f_y^{*2})]^{1/2}dx\,dy \quad . \tag{2}$$

We suppose that ε is small. Expanding both sides of (2) in powers of ε and observing again that $\underline{D}$ and $\underline{D}^*$ have the same area, we obtain, after obvious cancellation and division by $\varepsilon^2/2$, in letting ε tend to 0, that

$$\iint_{D} (f_x^2 + f_y^2)dx\,dy \geq \iint_{D^*} (f_x^{*2} + f_y^{*2})dx\,dy \quad . \tag{3}$$

IV. Steiner symmetrization diminishes the Dirichlet integral of a positive function vanishing along the boundary.

The foregoing derivation (21) is suggested by the usual introduction of Dirichlet's integral in the theory of vibrating membranes (26, vol. 1, p.307). The method, unfortunately, leaves little hope for discussing the sign of equality..

A 4. Applications. We derive several consequences from the fundamental result IV of sect. A 3.

(a) We consider a plane domain D, its principal frequency Λ and the function f satisfying 5.1 (1) for which equality is attained in 5.1 (3):

$$\Lambda^2 = \frac{\iint_{D} (f_x^2 + f_y^2)dx\,dy}{\iint_{D} f^2\,dx\,dy} \quad . \tag{1}$$

Steiner symmetrization (notation as in sect. A 1 - A 3) changes D into D^*, f into f^*, diminishes the numerator in (1), see A 3 (3), and leaves the denominator unchanged, see A 2 (3). Combining this with the inequality 5.1 (3) applied to D^* (instead of D), we obtain

$$\Lambda^2 \geq \frac{\iint_{D^*} (f_x^{*2} + f_y^{*2})dx\,dy}{\iint_{D^*} f^{*2}dx\,dy} \geq \Lambda^{*2} \quad , \tag{2}$$

and so $\Lambda \geq \Lambda^*$. That is, Steiner symmetrization diminishes the principal frequency.

(b) Applying A 3 (3) and A 2 (3) to 5.1 (2) in a similar manner, we obtain: Steiner symmetrization increases the torsional rigidity of a simply connected plane domain. The restriction to simply connected domains, unnecessary in the foregoing case, becomes necessary here, because the inequality characterizing P in the case of a multiply connected domain is more complicated than 5.1 (2); see 5.9A (c).

(c) Let now D be a multiply connected domain, the boundary of which consists of a finite number of smooth curves. Some of these curves constitute the entrance of D, others the exit of D, in the terminology of sect. 2.7A (a); there is no adiabatic portion. We assume that the outer boundary curve belongs to the entrance. Let f stand for an arbitrary (smooth) function defined in D which takes the value 0 along the entrance and the value 1 along the exit. Then the conductance c of D is characterized by the inequality (compare 2.7A (b))

$$\iint_{D} (f_x^2 + f_y^2)dx\, dy \geq 4\pi c \quad ; \tag{3}$$

the right hand side is the minimum of the left hand integral for all admissible f.

We consider D as a difference of two sets

$$D = E - F \quad ;$$

E is the connected domain the boundary of which is formed by the entrance of D; F consists of the interior of those boundary curves of D which belong to its exit. The domain E contains D; E may be multiply connected and F may be disconnected. We restrict now our consideration to that particular function f for which equality is attained in (3). This function f is harmonic in D and takes there only values between 0 and 1. We extend the definition of f to the whole domain E by giving it the value 1 in the points of F. Therefore, f is continuous (and sufficiently smooth) throughout the domain E and vanishes along its boundary. We could extend the integral in (3) over E (instead of D) without changing its value.

Let the Steiner symmetrization considered change D, E, F, f and c into D^*, E^*, F^*, f^* and c^*, respectively; E^*, F^* and f^* are defined by the symmetrization immediately. We define $D^* = E^* - F^*$. Let us observe that E^* is necessarily simply connected, but F^* may be disconnected; D^* is multiply connected. In defining c^*, the conductance of D^*, we consider as the entrance of D^* its outer boundary curve, the boundary of E^*, and as exit its inner boundary curves (or curve) which bound also F^*. The function f^* takes the value 0 along the entrance of D^* and 1 along its exit. The fundamental inequality A 3 (3), applied to the function f for which equality is attained in (3) in the manner used under (a), shows that $c \geq c^*$. That is, Steiner symmetrization diminishes the conductance, in absence of an adiabatic portion of the boundary.

We note two important limiting cases. In the first, the entrance degenerates into an infinitely large circle, in the second the exit degenerates into an infinitely small circle of center a. Considering these cases, we obtain (compare 2.3 (2) and (3)): Steiner symmetrization diminishes the outer radius $\bar{r}$ and increases the inner radius r_a.

The definition of the inner radius of a multiply connected domain indicated in 5.9A (d) and the foregoing proof apply without further remarks only to "regular" domains bounded by a finite number of smooth curves. Yet, considering an increasing sequence of such regular domains which exhausts a given domain D we can define r_a for D and prove the theorem stated in the most general case. (The definition of Steiner symmetrization given in sect. 7.1 is applicable to the most general case.) Similarly, instead of the outer radius $\bar{r}$ of a closed domain surrounded by a single curve we can consider the more general concept of the transfinite diameter due to Fekete (6). An appropriate process of exhaustion shows that Steiner symmetrization diminishes the transfinite diameter.

(d) From here on to the end of the present section (but not further) we call "quadrilateral" a plane domain D contained in an infinite strip S between two parallel lines, the distance of which is h, the width of S. The entry of D is a curve C_0 joining the two parallels, the exit is another curve C_1 of the same kind (see Fig. 4) and the adiabatic portion of the boundary consists of two segments, one on each of the parallels. With its boundary so divided, the quadrilateral D has a well defined conductance c; see sect. 2.7A (a). We shall prove that Steiner symmetrization with respect to a perpendicular to its strip diminishes the conductance of the quadrilateral.

Observe that in this statement, in opposition to the foregoing, the line of symmetrization is subject to a restriction. In order to prove our statement, we shift the quadrilateral D along its strip S into another position not overlapping with the original position and call the domain in the new position D'; see Fig. 4. Let us call E the domain consisting of

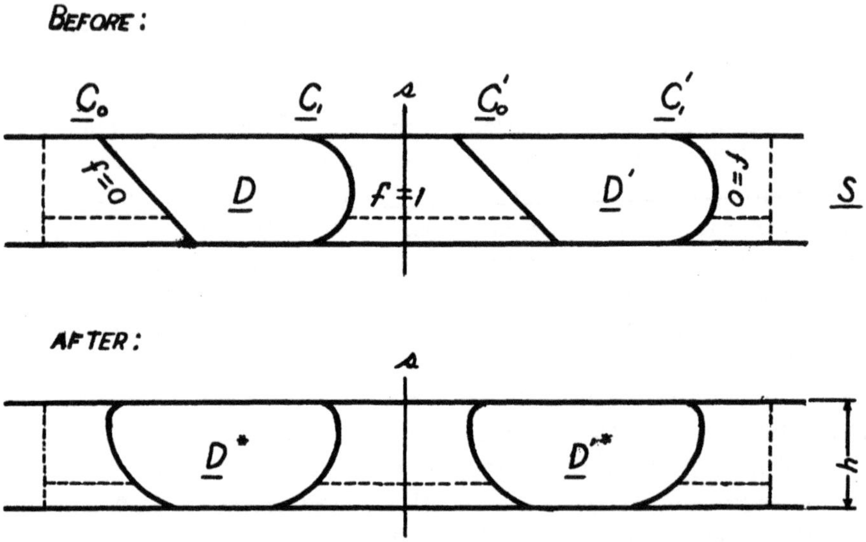

Fig. 4

$\underline{D}$, $\underline{D}'$ and the portion of $\underline{S}$ between them; in Fig. 4, $\underline{E}$ is bounded by the curves $\underline{C}_0$, $\underline{C}'_1$ and two rectilinear segments belonging to the boundary of $\underline{S}$. We define a function f in $\underline{E}$ as follows: $f = 0$ along the curves $\underline{C}_0$ and $\underline{C}'_1$, $f = 1$ along the curves $\underline{C}_1$ and $\underline{C}'_0$ and in that part of $\underline{E}$ (between the latter curves) which belongs neither to $\underline{D}$ nor to $\underline{D}'$. In $\underline{D}$ the function f coincides with the harmonic function that arises in the definition of the conductance c: it takes the values 0 and 1 along $\underline{C}_0$ and $\underline{C}_1$, respectively, as stated, and its normal derivative vanishes along the rectilinear (adiabatic) portion of the boundary of $\underline{D}$. The function is analogously defined in $\underline{D}'$; it coincides, in fact, with $1 - f$ as defined in $\underline{D}$, but shifted into $\underline{D}'$. Thus f is continuous throughout $\underline{E}$ and the integral of the square of its gradient over $\underline{E}$ (the Dirichlet integral) equals $8\pi c$. We introduce rectangular coordinates x, y, z so that $\underline{E}$ lies in the x,y-plane. We call $\underline{B}$ the solid bounded by $\underline{E}$ from below, by the surface $z = f(x,y)$ from above and by two planes, parallel to the z-axis and passing through the two parallels bounding $\underline{S}$, laterally.

Symmetrization with respect to a plane perpendicular to the parallels bounding $\underline{S}$ (which intersects the plane of Fig. 4 in the line s) changes $\underline{B}$, f, $\underline{D}$, $\underline{D}'$ and c into $\underline{B}^*$, f^*, $\underline{D}^*$, $\underline{D}'^*$ and c^*, respectively. It is easy to see that $\underline{D}^*$ itself can be considered as arising from $\underline{D}$ by Steiner symmetrization with respect to a line parallel to s; examine the dotted lines in Fig. 4. The surface surrounding $\underline{B}$ consists of four parts of which three (bounding $\underline{B}$ from below and laterally) are situated in three corresponding planes perpendicular to the plane of symmetrization; the areas of these three parts remain unchanged by symmetrization by theorem III of sect. A 1. Therefore, the fourth part must be diminished by symmetrization; this fourth part is expressed in terms of f similarly as the left hand side of A 3 (1) (but now the integral is extended over $\underline{E}$, or over $\underline{D}$ and $\underline{D}'$). We pass from the surface area to the Dirichlet integral and hence to the conductance (see A 3 (1), (2), and (3), and (3) of the present section; use the method under (a)) and we find that $c \geq c^*$, as asserted.

The same method yields other results, especially certain particular cases of the following fact which can be fully proved by the method of Note E: <u>the conductance of a quadrilateral situated in a given strip and having a given area is a minimum when the quadrilateral is a rectangle</u>. If we let h stand for the width of the strip (as before) and A for the area of the quadrilateral, the theorem just stated is equivalent to the inequality

$$c \geq \frac{h^2}{4\pi A} \quad . \tag{4}$$

A 5. <u>Schwarz symmetrization of the surface-area</u>. A (general) solid $\underline{B}$ and a solid of revolution $\underline{B}^*$, the boundary surface of which is simply connected, are in the following geometric relation: any plane that is perpendicular to the axis of revolution of $\underline{B}^*$ and intersects one of the solids $\underline{B}$

and B^* intersects also the other and the two intersections are equal in area. The intersection with B^* is a circular disk, that with B may be of any form. If B and the axis of revolution of B^* are given, B^* is uniquely determined. We say that B is transformed into B^* by symmetrization with respect to an axis (the axis of revolution of B^*) or Schwarz symmetrization. The Schwarz symmetrization has important properties, analogous to those of the Steiner symmetrization discussed in sect. A 1 and A 3.

I. Schwarz symmetrization leaves the volume unchanged.

II. Schwarz symmetrization diminishes the surface-area.

III. Schwarz symmetrization leaves unchanged the area of the intersection with any plane perpendicular to the axis of symmetrization.

IV. Schwarz symmetrization diminishes the Dirichlet integral of a positive function vanishing along the boundary.

The term "diminishes" means again "does not increase". The essential theorem is II; it has been proved by H. A. Schwarz.* Theorem IV follows from II by the method of sect. A 3, I is obvious (follows from "Cavalieri's principle"), and III is quite trivial: it merely restates the definition of Schwarz symmetrization. We confine ourselves to sketching an intuitive derivation of I, II, and III from the correspondingly numbered theorems of sect. A 1.**

We consider a solid B and two planes P_1 and P_2 which include an angle incommensurable with π ; we call the line of intersection of P_1 and P_2 the "axis a". We derive from B an infinite sequence of solids B_0, B_1, B_2, B_3, ... ; B_0 is identical with B, Steiner symmetrization with respect to P_1 changes B_{2n-2} into B_{2n-1} and Steiner symmetrization with respect to P_2 changes B_{2n-1} into B_{2n}, for $n = 1, 2, 3, \ldots$. By theorems I, II, and III of sect. A 1, each solid of the sequence has the same volume, its cross-section with a fixed plane perpendicular to the axis a has the same area, but the surface-area continually diminishes as n increases. If there is a limiting solid lim B_n, it must be symmetric both with respect to P_1 and with respect to P_2. Since these planes include an angle incommensurable with π, lim B_n must be a solid of revolution with axis a.*** As all solids of the sequence B_0, B_1, B_2, ... , the limiting solid lim B_n has an intersection with any plane perpendicular to the axis a which is equal in area to the intersection of the same plane with B. Therefore, Schwarz symmetrization with respect to the axis a transforms B into lim $B_n = B^*$. In view of the properties of the sequence, the volumes of lim B_n and B are equal and the

---

* H. A. Schwarz, Gesammelte mathematische Abhandlungen, vol.2,pp.327-335.

** The derivation can be made precise in the case of convex solids; see W. Blaschke, Kreis and Kugel, 1916, pp.86-90.

*** J. Steiner, Gesammelte Werke, 1882, vol.2, p.90.

surface-area of lim $\underline{B}_n$ is not superior to that of $\underline{B}$.

A 6. Applications. We consider a function f which is positive in a given domain of the x,y-plane, but vanishes along its boundary. Schwarz symmetrization changes f into a function f^* which is positive in a circular domain and vanishes along its periphery. The value of f^* at a point depends only on the distance of the point from the center of the circle and decreases (does not increase) as the distance increases. The theorem IV of the foregoing sect. A 5 is concerned with the passage from f to f^*; it is proved in a similar manner as theorem IV of sect. A 3 and has similar consequences.

(a) We take $\underline{D}$, Λ and f in the same meaning as in sect. A 4 (a), but we let now $\underline{D}^*$ stand for a circle with the same area as $\underline{D}$, Λ^* for the principal frequency of the circle $\underline{D}^*$ and f^* for the function defined in $\underline{D}^*$ into which f is transformed by Schwarz symmetrization. In this notation, theorem IV of sect. A 5 is expressed by A 3 (3). Also A 4 (1) and (2) remain valid and we can express the resulting inequality $\Lambda \geq \Lambda^*$ by saying that <u>of all domains with a given area the circle has the lowest principal frequency</u>. This is the theorem of Rayleigh, Faber, and Krahn, listed under No. 15 in the Table of sect. 1.21.

(b) In comparing (a) and (b) of sect. A 4 with (a) of the present section, we find by analogy (as the fourth term of a sort of proportion) that <u>of all simply connected cross-sections with a given area the circle has the highest torsional rigidity</u>. This is the theorem of de Saint-Venant, listed under No. 9 in the Table of sect. 1.21, first proved in <u>21</u> with the method of <u>24</u>, and proved quite differently in sect. 5.17.

(c) We adopt the notation of sect. A 4 (c) not fully, but in part. Let $\underline{D}$, $\underline{E}$, $\underline{F}$ and c have the same meaning as there, let the boundary of $\underline{D}$ be divided into entrance and exit as there, and let also f have the same meaning as in the latter part of sect. A 4 (c) so that equality holds in A 4 (3).

Let now $\underline{D}$, $\underline{E}$, $\underline{F}$, f and c be changed into $\underline{D}^*$, $\underline{E}^*$, $\underline{F}^*$, f^* and c^*, respectively, by Schwarz symmetrization (not by Steiner symmetrization). Therefore, $\underline{E}^*$ and $\underline{F}^*$ are circular disks with the same center, $\underline{E}^*$ contains $\underline{F}^*$, $\underline{D}^* = \underline{E}^* - \underline{F}^*$ is a circular ring, f^* depends only on the distance from the common center of $\underline{E}^*$ and $\underline{F}^*$, $f^* = 0$ on the boundary of $\underline{E}^*$, $f^* = 1$ both on the boundary of $\underline{F}^*$ and inside $\underline{F}^*$, and, finally, the Dirichlet integral of f^* over $\underline{D}^*$ is not superior to that of f over $\underline{D}$, by A 5 IV. Therefore, $c \geq c^*$, by A 4 (3) and the method of A 4 (a). Let $|\underline{E}|$, $|\underline{F}|$, $|\underline{E}^*|$, and $|\underline{F}^*|$ denote the areas of $\underline{E}$, $\underline{F}$, $\underline{E}^*$, and $\underline{F}^*$, respectively. Then $|\underline{E}^*| = |\underline{E}|$, $|\underline{F}^*| = |\underline{F}|$, by A 5 III. Evaluating c^*, the conductance of an annulus, we can write the inequality just derived in the form

$$(1) \qquad 1/c \leq \log(|\underline{E}|/|\underline{F}|) \quad .$$

That is, of all domains with a given area ($|E| - |F|$) and with given joint area of the holes bounded by the exit ($|F|$; the outer boundary curve of the domain belongs to the entrance) the ring bounded by two concentric circles has the minimum capacity. This theorem is due to Carleman in the particular case of a doubly connected (ring-shaped) domain; see 2. Carleman's method, based on conformal mapping, cannot be immediately extended to domains of higher connectivity, but the method of E 4 can be so extended.

We state two limiting cases, proved by quite different methods originally; see 18 and 29.

Of all bounded and closed sets with a given area the circular disk has the minimum transfinite diameter.

Of all open domains having a given area and containing a given point a the circle with center a has the maximum inner radius r_a.

These limiting cases correspond to those already considered in sect. A 4 (c) and, for doubly connected domains, in sect. 2.3. Since they do not impose any restriction on the connectivity, they contain much more than the second line of 1.12 (2).

(d) The theorem of de Saint-Venant, given under (b), deals only with the torsion of beams without cavities, that is, with simply connected cross-sections. The following theorem deals with the torsion of beams with cavities: Of all multiply connected cross-sections with given area and with given joint area of the holes, the ring bounded by two concentric circles has the maximum torsional rigidity.

We cannot bring here all details of the proof (see 25) but we wish to point out the role of Schwarz symmetrization and the differences from, and the agreements with, the foregoing case discussed under (c).

We adopt the notation of sect. 5.9A (c) and consider the second variational problem there considered which leads to P". (The domain E of sect. A 4 (c), reconsidered under the foregoing (c), coincides with the domain Δ of sect. 5.9A (c) only in the particular case where all inner boundary curves of D belong to the exit.) The function f considered in that variational problem is represented by a surface which we can call the "hill". The slopes of the "hill" start from the outer boundary curve of D at the level 0. The "hill" rises everywhere above its base, the domain Δ. The "hill" has several horizontal plateaux or tablelands (over the holes of D) each of which may be at a different level. (The function f considered under (c) has tablelands over the same holes of D, but some are fixed at the bottom level 0 and others at the top level 1.)

Schwarz symmetrization transforms our hill into a "round hill" which has ring-shaped plateaux and, in general, as many different plateaux as the "hill". An exception arises, however, if two or more plateaux of the "hill" happen to have the same elevation; in such a case, their transforms by Schwarz symmetrization merge into a single annular tableland of the "round hill". (The function f considered under (c) goes over by Schwarz symmetrization into an f^* represented by a round hill with a single circular plateau at the level 1 and, possibly, with a single annular plain at the level 0.)

The passage from the "hill" to the "round hill" diminishes the Dirichlet integral, by A 5 IV, and this yields the main conclusion in the proof of the theorem stated. It remains, however, to discuss the "round hill" with several tablelands and to show that a certain "round hill" with only one circular tableland yields the maximum of the torsional rigidity; see 25.

[(e) The two quantities compared under (d), P and c, are most closely related when the general situation described under A 4 (c) is specialized as follows: the entrance of $\underline{D}$ consists of the outer boundary curve alone and all the inner boundary curves belong to the exit. Under this condition, let us consider again the function f discussed in the latter half of A 4 (c); this f yields equality in A 4 (3), takes the value 1 in the domain $\underline{F}$ and values between 0 and 1 in the domain $\underline{D}$. Therefore, extending all the integrations over $\underline{E}$ and letting $|\underline{F}|$ denote the area of $\underline{F}$, we have:

(2).
$$\frac{\pi c}{|\underline{F}|^2} = \frac{\iint (f_x^2 + f_y^2)dx\,dy}{4|\underline{F}|^2} > \frac{\iint (f_x^2 + f_y^2)dx\,dy}{4(\iint f\,dx\,dy)^2} \geqq \frac{1}{P} ;$$

the last step is based on 5.1 (2) as extended in 5.9A (c) (in the notation used there, P = P"). We obtain so a connection between the torsional rigidity of a multiply connected domain and its conductivity:

(3) $$P > \pi^{-1}c^{-1}\,|\underline{F}|^2 .$$

The opposite inequality

(4) $$P < \pi^{-1}c^{-1}\,|\underline{E}|^2$$

is not generally true, but is certainly valid if the stress function of the torsion problem has no maximum in the interior of $\underline{D}$ as we can see by an argument which is very similar to (2) but goes in the opposite direction. The inequality (4) can be taken for granted in many cases of practical interest which, however, we cannot discuss here.]

A 7. Circular symmetrization of the surface-area. A straight line a called the axis, a halfplane $\underline{H}$, and two solids, $\underline{B}$ and $\underline{B}^*$, are connected by the following geometric relations: a is the edge bounding $\underline{H}$. Let c be any circle with center on the axis a and in a plane perpendicular to a; if c intersects one of the two solids $\underline{B}$ and $\underline{B}^*$, it intersects also the other and the two intersections are equal in length. The intersection with B may consist of several arcs, but the intersection with B^* consists of just one arc bisected by H; therefore, B^* is symmetric with respect to H. If the halfplane H is given, its edge a is, of course, determined. If B and H are given, B^* is uniquely determined. We say that B is transformed into B^* by circular symmetrization with respect to the halfplane H. (See 22.) If the edge of the halfplane goes to infinity, so that the halfplane is finally extended into a full plane, the circular symmetrization with respect to the halfplane becomes in the limit the Steiner symmetrization with respect to the full plane. The circular symmetrization has important properties, analogous to those of the Steiner and the Schwarz symmetrizations discussed in sect. A 1, A 3, and A 5.

I. Circular symmetrization leaves the volume unchanged.

II. Circular symmetrization diminishes the surface-area.

III. Circular symmetrization leaves unchanged the area of the intersection with any plane perpendicular to the edge of the plane of symmetrization.

IV. Circular symmetrization diminishes the Dirichlet integral of a positive function vanishing along the boundary.

I and III are obvious, IV follows from II by the method of sect. A 3. In order to show the essential theorem II, we introduce cylindrical coordinates r, φ, z, linked to the rectangular coordinates x, y, z by the relations $x = r\cos\varphi$, $y = r\sin\varphi$, $z = z$. We place the coordinate system so that the half plane of symmetrization is characterized by the equation $\varphi = 0$. Let the circle c mentioned above have m arcs in common with B the endpoints of which are

$$(r,\varphi_1,z) \text{ and } (r,\varphi_2,z), \;\dots\; (r,\varphi_{2m-1},z) \text{ and } (r,\varphi_{2m},z)$$

respectively, where

$$\varphi_1 > \varphi_2 > \varphi_3 > \dots > \varphi_{2m} > \varphi_1 - 2\pi \quad .$$

This circle c has just one arc in common with B^* the endpoints of which are $(r,-\varphi,z)$ and (r,φ,z) where

$$(1) \qquad 2\varphi = \varphi_1 - \varphi_2 + \varphi_3 - \varphi_4 + \dots + \varphi_{2m-1} - \varphi_{2m} \quad .$$

We let V and S denote the volume and the surface-area of B, respectively, V^* and S^* the corresponding quantities for B^*. We find

$$(2) \qquad V = \iint r(\varphi_1 - \varphi_2 + \varphi_3 - \varphi_4 + \dots + \varphi_{2m-1} - \varphi_{2m})\,dr\,dz \quad ,$$

$$(3) \qquad V^* = 2\iint r\,\varphi\,dr\,dz \quad ,$$

(4) $$S = \iint \sum_{\nu=1}^{2m} \left\{ 1 + r^2 \left[\left(\frac{\partial \varphi_\nu}{\partial r} \right)^2 + \left(\frac{\partial \varphi_\nu}{\partial z} \right)^2 \right] \right\}^{1/2} dr\, dz \quad ,$$

(5) $$S^* = 2 \iint \left\{ 1 + r^2 \left[\left(\frac{\partial \varphi}{\partial r} \right)^2 + \left(\frac{\partial \varphi}{\partial z} \right)^2 \right] \right\}^{1/2} dr\, dz \quad .$$

These four integrals have the same domain of integration, situated in the halfplane of symmetrization where r and z are rectangular coordinates. This domain consists of the points of intersection of the halfplane with the circles c intersecting $\underline{B}$ and $\underline{B}^*$. Of course, $m = m(r,z)$. Comparing (1), (2), and (3), we see that

(6) $$V = V^*$$

which is the property I. Comparing (1), (4), and (5) and observing that the sum of the lengths of the 2m vectors with components

$$1 \; , \; (-1)^{\nu-1}\, r \, \frac{\partial \varphi_\nu}{\partial r} \; , \; (-1)^{\nu-1}\, r \, \frac{\partial \varphi_\nu}{\partial z}$$

($\nu = 1,2,\ldots,2m$) is not less than the length of their sum, we find that

(7) $$S \geq S^*$$

which proves property II. We do not enter into the proof of III, but observe that more is true: the intersections of $\underline{B}$ and $\underline{B}^*$ with any surface of revolution about the edge of the halfplane of symmetrization as axis are equal in area.

A 8. Applications. In rectangular coordinates x, y, z, the half-plane of symmetrization considered in the foregoing sect. A 7 is characterized by $x > 0$, $y = 0$ and its edge by $x = 0$, $y = 0$. We consider a domain $\underline{D}$ in the x,y-plane, a function $f(x,y)$ which is positive inside $\underline{D}$ and vanishes along its boundary, and the solid $\underline{B}$ contained between the surface $z = f(x,y)$ and the domain $\underline{D}$. Circular symmetrization changes $\underline{B}$, $\underline{D}$ and f into $\underline{B}^*$, $\underline{D}^*$ and f^*, respectively, where $z = f^*(x,y)$ is the equation of the surface which, together with $\underline{D}^*$, forms the boundary of the solid $\underline{B}^*$. For a fixed positive r and $|\varphi| \leq \pi$, the function $f^*(r\cos\varphi, r\sin\varphi)$ is an even function of φ which decreases (not necessarily strictly) as $|\varphi|$ increases. Theorem IV of sect. A 7 is concerned with the functions f and f^*, and the domains $\underline{D}$ and $\underline{D}^*$; it is expressed by A 3 (3), proved in the same fashion as the theorems numbered IV in sect. A 3 and A 5, and applied similarly. Referring to sect. A 4 and A 6 for the method of derivation, we just list the results and some of their further consequences.

(a) Circular symmetrization diminishes the principal frequency. We illustrate this theorem by an application. We consider a circle with center

O and radius a, a given point at the distance d from O where $d < a$, a simply connected domain D which is contained in the circle but does not contain the given point in its interior, and, finally, the simplest domain D_1 of the same kind as D: the boundary of D_1 consists of the circle itself and of a rectilinear slit of length $a - d$ joining radially the given point to the periphery. We prove that the principal frequency of D is not lower than that of D_1. In fact, circular symmetrization of D with respect to a suitable halfplane the edge of which passes through O yields a domain D^* which does not contain in its interior the slit described above. Therefore, D^* is contained in D_1 and so the principal frequency of D_1 is not higher than that of D^* which, owing to circular symmetrization, is not higher than that of D.

(b) Circular symmetrization increases the torsional rigidity. It follows that of all domains D described under (a) the domain D_1 has the maximum torsional rigidity.

(c) Let D, E, F, f and c (the meaning is the same as in the latter part of sect. A 4 (c) and in sect. A 6 (c)) be changed into D^*, E^*, F^*, f^* and c^*, respectively, by circular symmetrization (not by Steiner or Schwarz symmetrization). Again $c \geq c^*$, that is, circular symmetrization diminishes the conductance. This fact has many interesting consequences, some of which we are going to discuss.

(d) We are given four numbers a_1, a_2, a_3, and a_4 subject to the conditions

$$0 \leqq a_1 \ , \ a_1 \leqq a_2 < a_4 \ , \ a_1 < a_3 \leqq a_4 \ .$$

We consider circles (that is, circular lines) in the x,y-plane around the common center O (the point $x = 0$, $y = 0$) and let $[r]$ denote the circle with radius r. The open domain D (the boundary of which consists again of entrance and exit) satisfies the following conditions:

D is contained in the ring between the circles $[a_1]$ and $[a_4]$;

if $r > a_3$ and there is a point of D on the circle $[r]$, there is on it also a point of the entrance of D;

if $r < a_2$ and there is a point of D on the circle $[r]$, there is on it also a point of the exit of D.

A simple example of such a domain is D_2, defined as follows: the entrance of D_2 consists of the circle $[a_4]$ and the line-segment characterized by

$$-a_4 < x \leqq -a_3 \ , \ y = 0 \ , \tag{1}$$

and the exit of D_2 consists of the circle $[a_1]$ and the line-segment characterized by

$$a_1 < x \leqq a_2 \ , \ y = 0 \ . \tag{2}$$

Let c and c_2 denote the conductances of $\underline{D}$ and $\underline{D}_2$, respectively. We shall prove that

$$(3) \qquad c \geqq c_2 \quad .$$

Preliminary remark. As is well known, the principal frequency Λ and the torsional rigidity P are monotonic domain functions, the former decreasing, the latter increasing. We used this under (a) and (b); the corresponding step in the following argument is based on the behavior of the conductance c, which is more complex. Let the domain $\underline{D}$ be contained in the domain $\underline{D}'$, but so that of the three portions of the boundary that each domain has (namely entrance, exit, and adiabatic portion) two portions are exactly the same for both domains. That is, the change of only one portion is responsible for expanding $\underline{D}$ into $\underline{D}'$. Let c and c' denote the conductance of $\underline{D}$ and $\underline{D}'$, respectively. If $\underline{D}$ expands into $\underline{D}'$ by the change of the entrance or the exit, we have $c \geqq c'$, but if $\underline{D}$ expands into $\underline{D}'$ by the change of the adiabatic portion, we have $c \leqq c'$.

In order to prove one part of this assertion, let f be the function for which equality is attained in A 4 (3). Let $\underline{D}$ expand into $\underline{D}'$ by modifying an arc of the entrance and extend the definition of f to $\underline{D}'$ by setting f = 0 at the points of $\underline{D}'$ which do not belong to $\underline{D}$. The function f so defined is continuous (sufficiently smooth) in $\underline{D}'$ and can be used to estimate c'. The integral in A 4 (3) does not change its value if it is extended over $\underline{D}'$ (instead of $\underline{D}$) provided that f is used in the sense just defined. So extended, however, the integral yields an upper bound for $4\pi c'$ and hence $c \geqq c'$. The other parts of our assertion can be similarly proved, and so can be proved the monotonicity of P and Λ, on the basis of the corresponding variational definitions 5.1 (2) and (3). The case of equality is not difficult to discuss, but we do not need it now and leave it aside.

We return to the proof of (3). Circular symmetrization with respect to the halfplane specified at the beginning of this chapter (its edges passes through 0) changes $\underline{D}$ into $\underline{D}^*$ and c into c^* with

$$(4) \qquad c \geqq c^*$$

by IV of sect. A 7. It follows from the three conditions imposed upon $\underline{D}$ that $\underline{D}^*$ is contained between the circles $[a_1]$ and $[a_4]$; no point of the segment (1) belongs to the interior of $\underline{D}^*$ and, if a point of (1) belongs to the boundary of $\underline{D}^*$, it belongs to the entrance of $\underline{D}^*$ and $f^* = 0$ at the point; no point of the segment (2) belongs to the interior of $\underline{D}^*$ and if a point of (2) belongs to the boundary of $\underline{D}^*$ it belongs to the exit of $\underline{D}^*$ and $f^* = 1$ at the point. In fact, the whole segment (2) and the circular disk bounded by $[a_1]$ must belong to the closed set $\underline{F}^*$. Therefore, $\underline{D}^*$ is contained in $\underline{D}_2$. In fact, $\underline{D}^*$ can be expanded into $\underline{D}_2$ by two subsequent changes: by changing the entrance first and the exit afterwards. Therefore, by our preliminary remark

$$(5) \qquad c^* \geqq c_2$$

and (3) results from (4) and (5). We mention a few particular cases and limiting cases of the result just proved.

(I) $a_1 = a_2 = 0$, $a_3 = d$, $a_4 = \infty$. Taking as the exit of $\underline{D}$ a circle $[\varepsilon]$ and letting ε tend to 0, we find: Let r_0 be the inner radius with respect to 0 of a domain containing the point 0. Let the domain be such that no circle $[r]$ with $r > d$ lies fully in the interior of $\underline{D}$. Then

$$(6) \qquad r_0 \leqq 4d \quad .$$

An essential feature of this theorem is that no restriction is imposed upon the connectivity of the domain considered. If the domain is supposed to be simply connected, (6) reduces to a well known part of the classical distortion theorem foreseen by Koebe.*

(II) $a_1 = a_2 = 0$, $a_3 = d$, $a_4 = a$. The domain with inner radius r_0 considered under (I) is subject now to the further condition that it is contained in the interior of the circle $[a]$. Then we have the inequality

$$(7) \qquad r_0 \leqq \frac{4a^2d}{(a + d)^2} \quad ,$$

which is sharper than (6). The right hand side of (7) is the inner radius, with respect to 0, of the domain $\underline{D}_1$, considered under (a)**; under the present conditions for a_1, a_2, a_3 and a_4, $\underline{D}_2$ becomes $\underline{D}_1$ punctured at the point 0. Again, no restriction regarding connectivity is needed for (7). In the particular case of a simply connected domain the result is due to Pick.***

(III) $a_1 = a_2 > 0$, a_3 finite, $a_4 = \infty$. The particular case in which $\underline{D}$ is doubly connected is due to Grötzsch.****

---

* See e.g., G. Pólya and G. Szegö, Aufgaben und Lehrsätze, vol. 2, p. 26, problem 147.

** Pólya and Szegö, l.c., p. 19, problem 108.

*** G. Pick, Über die konforme Abbildung eines Kreises auf ein schlichtes und zugleich beschränktes Gebiet, Wiener Berichte, Abt. IIa, vol. 126 (1917) pp. 247-263.

**** H. Grötzsch, Über einige Extremalprobleme der konformen Abbildung, Berichte d. math.-physik. Klasse, Sächsische Akademie d. Wissenschaften, Leipzig, vol. 80 (1928) pp. 367-376.

(IV) $0 = a_1 < a_2, a_3$ finite, $a_4 = \infty$. The particular case in which D is doubly connected is due to Teichmüller.*

Other important applications of circular symmetrization to the theory of analytic functions have been found** and, perhaps, still others may be expected.

---

* O. Teichmüller, Untersuchungen über konforme und quasikonforme Abbildung, Deutsche Mathematik, vol. 3 (1938), pp. 621-678.

** W. K. Hayman, Symmetrization in the Theory of Functions, Technical Report No. 11, Office of Naval Research, Stanford University, August 31, 1950. See also a paper by V. Wolontis, The change of resistance under circular symmetrization, presented to Section II of the International Congress of Mathematicians, Cambridge, 1950.

NOTE B. ON CONTINUOUS SYMMETRIZATION*

B 1. Problem. (a) Let C be a plane curve and ℓ an arbitrary straight line in the same plane. Steiner symmetrization replaces the curve C by another curve C^* which is symmetrical with respect to ℓ. In the transition from C to C^* various quantities associated with C change in a predictable manner. For instance:

(α) The area of the domain D enclosed by C is unchanged.

(β) The length of C is diminished.

(γ) The outer radius (transfinite diameter) of C is diminished.

(δ) The inner radius of C with respect to a fixed point on ℓ (which in this case must intersect C) is increased.

(b) We deal here with the following question. Is it possible to define a transformation T_λ depending in a continuous way on the parameter λ, $0 \leqq \lambda \leqq 1$, such that the following conditions are satisfied:

I. T_0 is the identity.

II. T_1 is the transformation replacing C by the symmetrized curve C^*, that is, T_1 is Steiner's symmetrization with respect to the line ℓ.

III. For any λ, $0 \leqq \lambda \leqq 1$, T_λ has the same effect as described under (α) - (δ).

We denote by $C^{(\lambda)}$ the curve arising from C by the transformation T_λ so that

$$C^{(0)} = C, \quad C^{(1)} = C^* \quad .$$

The transformation T_λ depends, of course, on the line ℓ.

We shall give an answer to the previous question in the important special case when the curve C is convex. Conditions (α) and (β) will be satisfied also in the more general case when no line perpendicular to ℓ intersects the curve C in more than two points.

B 2. Continuous symmetrization in a plane. (a) Let C be a given curve, ℓ a given straight line. We suppose that no line perpendicular to

---

* Professor M. Shiffman collaborated on this subject.

ℓ has more than two points in common with $\underline{C}$.

We denote by a and b the end-points of such a segment of intersection, indicating by the same letters the ordinates of these points in a cartesian coordinate system x, y, the x-axis of which is ℓ ; we assume that $a > b$.

Let $0 \leq \lambda \leq 1$. We shift the segment a,b along its line into the position a', b' such that

$$(1) \qquad T_\lambda : \quad a' = a - \lambda \frac{a+b}{2} \quad , \quad b' = b - \lambda \frac{a+b}{2} \quad .$$

That is, the segment a,b is shifted with uniform velocity in the position where it is bisected by the x-axis (the line ℓ). This is the transformation T_λ we want to consider. It may be called <u>continuous</u> <u>symmetrization</u>. For $\lambda = 0$ it is the identity and for $\lambda = 1$ it results in complete symmetrization.

If we move ℓ into another position ℓ' parallel to ℓ , continuous symmetrization with respect to ℓ and ℓ' (with the same λ) results in two congruent curves; they arise from each other by shifting by λd where d is the distance of ℓ and ℓ'.

(b) The transformation T_λ leaves the area of the domain $\underline{D}$ bounded by $\underline{C}$ unchanged. This is obvious.

We show now that the length of $\underline{C}^{(\lambda)}$ is steadily decreasing (never increasing) as λ increases from 0 to 1. For this purpose, we denote the "upper arc" forming the curve $\underline{C}$ by $y = y_1(x)$ and the "lower arc" by $y = y_2(x)$, $y_1 > y_2$. Then the arc elements at a and b are given by

$$(2) \qquad ds_1^2 = dx^2 + dy_1^2 \ , \quad ds_2^2 = dx^2 + dy_2^2 \quad .$$

The transformation T_λ changes the ordinates y_1 and y_2 into

$$(3) \qquad y_1^{(\lambda)} = y_1 - \lambda(y_1 + y_2)/2, \quad y_2^{(\lambda)} = y_2 - \lambda(y_1 + y_2)/2$$

and so ds_1^2 and ds_2^2 into

$$(4) \qquad \begin{aligned} (ds_1^{(\lambda)})^2 &= dx^2 + [dy_1 - \lambda(dy_1 + dy_2)/2]^2 \quad , \\ (ds_2^{(\lambda)})^2 &= dx^2 + [dy_2 - \lambda(dy_1 + dy_2)/2]^2 \quad . \end{aligned}$$

We set

$$(5) \qquad \frac{dy_1}{dx} = p_1 \ , \quad \frac{dy_2}{dx} = p_2$$

and

$$(6) \qquad ds_1^{(\lambda)} + ds_2^{(\lambda)} = f(\lambda)dx$$

where

$$f(\lambda) = \{1 + [p_1 - \lambda(p_1 + p_2)/2]^2\}^{1/2}$$

(7)
$$+ \{1 + [p_2 - \lambda(p_1 + p_2)/2]^2\}^{1/2}$$

Note that $\underline{C}^{(\lambda)}$ is obtained by integrating (6) and that

(8)
$$f(2 - \lambda) = f(\lambda) \quad .$$

(c) Let A, B and C be real, $A > 0$, $AC - B^2 > 0$ and

$$\varphi(\lambda) = (A\lambda^2 + 2B\lambda + C)^{1/2} \quad .$$

Then

$$\varphi''(\lambda) = (AC - B^2)\,[\varphi(\lambda)]^{-3}$$

and so the function $\varphi(\lambda)$ is convex from below. In view of this fact and (7), $f(\lambda)$ is convex from below. We infer from (8) that $f(\lambda)$ is steadily decreasing as λ increases from 0 to 1, and so does (6), and so does the total length of $\underline{C}^{(\lambda)}$.

B 3: Continuous symmetrization in space. The definition of T_λ is analogous in space. It applies to a given surface $\underline{A}$ and refers to a given plane p; $0 \leq \lambda \leq 1$. It changes $\underline{A}$ into $\underline{A}^{(\lambda)}$. We assume again that no line perpendicular to p has more than two points in common with $\underline{A}$. The volume of the domain $\underline{B}$ bounded by $\underline{A}$ remains unchanged under the transformation T_λ .

The surface area of $\underline{A}^{(\lambda)}$ is steadily decreasing as λ increases from 0 to 1. We have to show that an element of $\underline{A}^{(\lambda)}$ which is an expression of the form

$$\{1 + [p_1 - \lambda(p_1 + p_2)/2]^2 + [q_1 - \lambda(q_1 + q_2)/2]^2\}^{1/2}$$

$$+ \{1 + [p_2 - \lambda(p_1 + p_2)/2]^2 + [q_2 - \lambda(q_1 + q_2)/2]^2\}^{1/2}$$

behaves so, and this is clear after the foregoing B 2 (c).

B 4. Capacity. (a) Let $\underline{C}_0$ and $\underline{C}_1$ be two curves in the plane, $\underline{C}_0$ in the interior of $\underline{C}_1$. We consider all functions u defined in the ring-shaped domain $\underline{D}$ bounded by $\underline{C}_0$ and $\underline{C}_1$ and satisfying the boundary conditions

(1)
$$u = 0 \text{ on } \underline{C}_0 \, , \quad u = 1 \text{ on } \underline{C}_1 \quad .$$

The minimum $4\pi c$ of the integral

$$\iint |\text{grad } u|^2 \, d\sigma \tag{2}$$

where the integration is extended over $\underline{D}$ ($d\sigma$ is the area-element) defines the capacity c of the domain $\underline{D}$; see 2.3.

Let $\underline{C}_0$ and $\underline{C}_1$ be convex curves and let ℓ be an arbitrary line. We apply the continuous symmetrization T_λ, $0 \leq \lambda \leq 1$, to the curves $\underline{C}_0$ and $\underline{C}_1$ with respect to ℓ, resulting in certain curves $\underline{C}_0^{(\lambda)}$ and $\underline{C}_1^{(\lambda)}$. Obviously $\underline{C}_0^{(\lambda)}$ is contained in $C_1^{(\lambda)}$. We prove that the capacity of the domain $\underline{D}$ between $\underline{C}_0$ and $\underline{C}_1$ is diminished (not increased) by this transformation; i.e., if we denote by $c(\lambda)$ the capacity of the ring-shaped domain $\underline{D}^{(\lambda)}$ bounded by $\underline{C}_0^{(\lambda)}$ and $\underline{C}_1^{(\lambda)}$, we have

$$c(\lambda) \leq c \quad . \tag{3}$$

Of course, it follows also that $c(\lambda)$ is a decreasing function of λ if $\lambda \geq 0$.

(b) For the proof we observe first that the domain $\underline{D}$ can be mapped conformally onto a domain bounded by two concentric circles k_0 and k_1. The level curves $\underline{C}_\rho$ of the minimizing function u of Dirichlet's integral correspond to the concentric circles k_ρ between k_0 and k_1. The convexity of $\underline{C}_\rho$ is equivalent with the positivity of a certain harmonic function which is regular between k_0 and k_1.* Now if a harmonic function is positive on the boundary of a certain domain it must remain positive in the interior. Hence <u>all level curves C_ρ must be convex</u>.

We apply now the transformation T_λ to $\underline{C}_\rho$ and denote the resulting curve by $\underline{C}_\rho^{(\lambda)}$. We define $\bar{u} = \rho$ on $\underline{C}_\rho^{(\lambda)}$. We can prove then that this transformation diminishes Dirichlet's integral (2). Indeed, this is a consequence of the fact that the transformation T_λ diminishes the surface area. See sect. A 3.

This establishes the assertion.

(c) Let $\underline{C}$ be a convex curve, $\bar{r}$ its outer radius (transfinite diameter). From the previous result we conclude that <u>the outer radius is diminished (not increased)</u> by the transformation T_λ, $0 \leq \lambda \leq 1$.

Indeed, let us consider the ring-shaped domain bounded by the curve $\underline{C}$ and by a circle Γ of radius ω about the origin. Let $c_1 = c_1(\omega)$ be the capacity of this domain. Then we have 2.3 (3). Applying T_λ to $\underline{C}$ a new curve $\underline{C}^{(\lambda)}$ arises while Γ is only shifted by T_λ. Thus the assertion follows from the previous result.

We deal in a similar fashion with the inner radius.

B 5. <u>Example</u>. We illustrate our result by the following application: <u>Consider all parallelograms with given base and height. The outer</u>

---

* Cf. for instance, G. Pólya-G. Szegö, Aufgaben und Lehrsätze, vol. 1, p. 105, problem 108.

radius $\bar{r}$ monotonically decreases and the inner radius $\dot{r}$ with respect to the
center monotonically increases if the acute angle of the parallelogram is
increasing from 0 to $\pi/2$.

Indeed, continuous symmetrization of the parallelogram with respect to
a line parallel to the height results in a parallelogram of the same base
and height with a larger acute angle.

B 6. Polar moment of inertia. Let $\underline{D}$ be an arbitrary plane domain and
ℓ an arbitrary straight line through the centroid of $\underline{D}$. Continuous symmetrization of $\underline{D}$ with respect to ℓ diminishes the polar moment of inertia of $\underline{D}$
with respect to the centroid.

Following the notation of B 2 we denote by a and b the ordinates of two
points of $\underline{D}$ with the same abscissa in a cartesian coordinate system x,y,
the x-axis of which is ℓ; a > b. It suffices to show that

$$a^3 - b^3 > [a - \lambda(a + b)/2]^3 - [b - \lambda(a + b)/2]^3 .$$

Subtracting the right-hand expression from the left-hand one we obtain

$$\frac{3}{4} \lambda (2 - \lambda) (a - b) (a + b)^2 > 0 .$$

NOTE C. ON SPHERICAL SYMMETRIZATION

C 1. Definition. We consider a domain $\underline{B}$ in space and a domain $\underline{D}$ in a plane. In the foregoing (sect. 1.7-1.10, ch. 7, note A) we discussed various kinds of symmetrizations.

(a) Symmetrization of a space domain $\underline{B}$ with respect to a plane (Steiner symmetrization).

(b) Symmetrization of a plane domain $\underline{D}$ with respect to a line in the plane of $\underline{D}$ (Steiner symmetrization).

(c) Symmetrization of a space domain $\underline{B}$ with respect to a line (Schwarz symmetrization).

(d) Symmetrization of a plane domain $\underline{D}$ with respect to a center and a half-line issued from this center (circular symmetrization).

Now we consider a further kind of symmetrization which we call spherical symmetrization. This is the symmetrization of a space domain $\underline{B}$ with respect to a center O and a half-line ℓ issued from this center. The procedure is as follows. We intersect $\underline{B}$ by a sphere about O and replace the spherical domain of intersection (or the sum of the domains of intersections) by a single spherical cap of equal area on this sphere whose center is on the half-line ℓ.

Obviously the spherical symmetrization is the space analogue of the circular symmetrization. Also, Schwarz symmetrization (c) is a limiting case of the spherical symmetrization when the center tends to infinity along the fixed half-line which is the extension of the given half-line ℓ beyond the given center O.

Our main purpose is to investigate the effect of this symmetrization on the capacity of a given solid $\underline{B}$.

C 2. A theorem on capacity. We consider in a more general fashion a "hollow" solid $\underline{B}$ in space bounded by the closed surfaces $\underline{A}_0$ and $\underline{A}_1$ where $\underline{A}_0$ is in the interior of $\underline{A}_1$. We form Dirichlet's integral

$$D(u) = \iiint |\text{grad } u|^2 \, d\tau \tag{1}$$

where u is an arbitrary function defined in $\underline{B}$, satisfying the boundary conditions

(2) $$u = 0 \text{ on } \underline{A}_0 \ , \quad u = 1 \text{ on } \underline{A}_1$$

The volume element is denoted by $d\tau$. The capacity C of the ring-shaped solid $\underline{B}$ (condenser) has been defined in 2.2 as the minimum of $(4\pi)^{-1} D(u)$ for all functions u admitted above.

In 24 (see sect. 7.3) we proved that the capacity C is diminished by the Steiner symmetrization, C 1 (a), and by the Schwarz symmetrization C 1 (c). More precisely, the symmetrization of $\underline{B}$ is meant as follows. We symmetrize the solid bounded by $\underline{A}_0$ and call the boundary of the symmetrized solid $\bar{A}_0$; similarly we define $\bar{A}_1$. The symmetrized "hollow" solid $\bar{B}$ is bounded by $\bar{A}_0$ and $\bar{A}_1$.

We shall prove that <u>the capacity is diminished under the spherical symmetrization</u> described above.

We cite in particular the case when the interior surface $\underline{A}_0$ is a given surface $\underline{A}$ and the exterior surface $\underline{A}_1$ degenerates into the infinitely large sphere. The integral (1) must be extended then over the whole exterior of $\underline{A}$, and the minimum defines the capacity of $\underline{A}$ (with respect to the infinitely large sphere).

An interesting application (foreseen in sect. 7.10) is the following:

<u>Of all spherical domains with a given area on a given sphere, the spherical cap has the minimum capacity</u> (<u>with respect to the infinitely large sphere</u>).

C 3. <u>Preparations for the proof</u>. The proof follows the pattern of sect. 3.2. Let $\underline{A}_\rho$ be the level surfaces of the minimizing function u of C 2 (1), $0 \leq \rho \leq 1$. We subject the domain which is the closed interior of $\underline{A}_\rho$ to spherical symmetrization. The resulting domain has axial symmetry with respect to ℓ; every sphere about 0 intersects it in a cap with center on ℓ. We denote the boundary of the symmetrized domain by $\bar{A}_\rho$. The given domain $\underline{B}$ is transformed into $\bar{B}$ which is a domain of revolution about the axis ℓ, bounded by $\bar{A}_0$ and $\bar{A}_1$.

We define now $\bar{u} = \rho$ on $\bar{A}_\rho$ and intend to show that Dirichlet's integral is diminished in passing from the given domain $\underline{B}$ to the symmetrized domain $\bar{B}$ and from the function u to the "symmetrized" function $\bar{u}$. Obviously $\bar{u}$ is admissible for the domain $\bar{B}$, that is, it satisfies boundary conditions corresponding to C 2 (2). By such an argument we will arrive at the conclusion that the minimum of Dirichlet's integral is also diminished, and this will prove the assertion.

Let $r > 0$ and $0 < \rho < 1$. The sphere of radius r about the center intersects $\underline{A}_\rho$ in a set $s(r, \rho)$ which is in general a spherical curve or a system of such curves. We shall prove that not only the total Dirichlet integral is diminished but even each of its infinitesimal parts corresponding to the sum of the elements along the curve (or system of curves) $s(r,\rho)$. We let $A(r,\rho)$ denote that part of the spherical surface with radius r which lies inside $\underline{A}_\rho$. The curve (or system of curves) $s(r,\rho)$ forms the boundary of $A(r,\rho)$. We shall sometimes refer to $A(r,\rho)$ as "the spherical area included by $s(r,\rho)$." The intersection of the sphere mentioned with the closed interior of the symmetrized surface $\bar{A}_\rho$ will be a spherical cap of area $A(r,\rho)$ having the particular position described above.

(We may disregard here the degenerate case when the surface $\underline{A}_\rho$ has a piece of surface in common with the sphere of radius r about the center O. In such a case the set $s(r,\rho)$ will be a spherical surface rather than a curve.)

C 4. First part of the proof. We represent u in polar coordinates r, θ, φ, as $u = u(r,\theta,\varphi)$, r being the distance from O and θ the spherical distance on the sphere of radius r about O measured from the pole situated on the half-line ℓ; hence $\theta = 0$ characterizes ℓ. We denote by $\vec{t}$, $\vec{t}_1$, $\vec{t}_2$ unit vectors at an arbitrary point r, θ, φ pointing in the directions of the respective coordinate-lines. We have then

(1) $$\operatorname{grad} u = \frac{\partial u}{\partial r}\vec{t} + \frac{1}{r}\frac{\partial u}{\partial \theta}\vec{t}_1 + \frac{1}{r\sin\theta}\frac{\partial u}{\partial \varphi}\vec{t}_2$$
$$= f\vec{t} + g\vec{t}_1 + h\vec{t}_2 \quad .$$

Here $f\vec{t}$ is the component of the gradient perpendicular to the sphere of radius r about O, while $g\vec{t}_1 + h\vec{t}_2$ is the component of the gradient on the sphere.

(a) Our first concern is to compare the spherical curve $s(r,\rho)$ with the central projection s' of $s(r + dr, \rho)$ onto the sphere r; $dr > 0$. The difference of the areas included by $s(r,\rho)$ and s', respectively, is

(2) $$\left|A(r,\rho) - \frac{r^2}{(r+dr)^2}A(r+dr,\rho)\right| = r^2\left|\frac{A(r,\rho)}{r^2} - \frac{A(r+dr,\rho)}{(r+dr)^2}\right|$$
$$= r^2 dr\left|\frac{d}{dr}\frac{A(r,\rho)}{r^2}\right| = dr\left|\frac{2A}{r} - A_r\right|$$

where A_r denotes the partial derivative of $A(r,\rho)$ with respect to r.

We determine the scalar η in such a way that the vector $\eta(g\vec{t}_1 + h\vec{t}_2)$ (which is perpendicular to $s(r,\rho)$) ends on the projected curve s'. Then the vector

(3) $$dr.\vec{t} + \eta(g\vec{t}_1 + h\vec{t}_2)$$

lies in the surface $u = \rho$, hence it must be perpendicular to grad u. This fact enables us to compute η. Forming the scalar product of (3) and (1),

$$f\,dr + \eta\,(g^2 + h^2) = 0\,, \qquad \eta = -f(g^2 + h^2)^{-1}\,dr \tag{4}$$

results. This yields for the piece of the normal of $s(r,\rho)$ ending on s':

$$|\eta|\,(g^2 + h^2)^{1/2} = |f|\,(g^2 + h^2)^{-1/2}\,dr \;. \tag{5}$$

Thus we obtain

$$|2A/r - A_r| \leqq \int_{s(r,\rho)} |f|\,(g^2 + h^2)^{-1/2}\,ds \tag{6}$$

where the integration is extended over the spherical curve $s(r,\rho)$.

(b) On the other hand we compare the spherical curves $s(r,\rho)$ and $s(r,\rho + d\rho)$ where $d\rho > 0$. For this purpose we determine the piece dn of the normal of $s(r,\rho)$ ending on $s(r,\rho + d\rho)$. Since the spherical component of the gradient is $g\vec{t}_1 + h\vec{t}_2$ we have

$$(g^2 + h^2)^{1/2} = d\rho/dn \;. \tag{7}$$

This yields for the area between $s(r,\rho)$ and $s(r,\rho + d\rho)$:

$$\begin{aligned} A(r,\rho + d\rho) - A(r,\rho) &= A_\rho\, d\rho \\ &= \int_{s(r,\rho)} ds.dn = d\rho \int_{s(r,\rho)} (g^2 + h^2)^{-1/2}\,ds \end{aligned} \tag{8}$$

where A_ρ is* the partial derivative of $A(r,\rho)$ with respect to ρ. Hence

$$A_\rho = \int_{s(r,\rho)} (g^2 + h^2)^{-1/2}\,ds \;. \tag{9}$$

(c) Finally we consider the part of Dirichlet's integral D(u) corresponding to the infinitesimal domain between the spheres r and r + dr and between the level surfaces $\underline{A}_\rho$ and $\underline{A}_{\rho + d\rho}$; $dr > 0$, $d\rho > 0$. We have

$$\begin{aligned} &|\text{grad } u|^2\, dr.\, r^2 \sin\theta\, d\theta\, d\varphi \\ &= (f^2 + g^2 + h^2)\, dr.\, r^2 \sin\theta\, d\theta\, d\varphi \end{aligned} \tag{10}$$

* The reader should distinguish carefully between A_ρ and $\underline{A}_\rho$ and $\bar{A}_\rho$.

the last factor being the area-element on the sphere of radius r. This element is identical with the element of the spherical domain between the curves $s(r, \rho)$ and $s(r, \rho + d\rho)$. For this we have obtained in (b):

$$ds.dn = (g^2 + h^2)^{-1/2} \, d\rho \; ds. \tag{11}$$

Hence the part of the Dirichlet integral we are interested in will be

$$\begin{aligned} &\int_{s(r,\rho)} (f^2 + g^2 + h^2)\, dr.\, (g^2 + h^2)^{-1/2} \, d\rho \;\; ds \\ &= dr\, d\rho \int_{s(r,\rho)} f^2 (g^2 + h^2)^{-1/2} \, ds + dr\, d\rho \int_{s(r,\rho)} (g^2 + h^2)^{1/2} \, ds \end{aligned} \tag{12}$$

C 5. <u>Second part of the proof</u>. (a) We form now the corresponding part of Dirichlet's integral in the symmetrized situation assuming again that $dr > 0$, $d\rho > 0$. The function $\bar{u}$ is defined by the condition that $\bar{u} = \rho$ on the symmetrized surface $\bar{A}_\rho$. The intersection of $\bar{A}_\rho$ with the sphere r about 0 is a cap with center on ℓ ; denoting the angle of this cap by θ its area will be $4\pi r^2 \sin^2 (\theta/2)$, so that according to definition of the spherical symmetrization

$$A(r, \rho) = 4\pi r^2 \sin^2 (\theta/2) \quad . \tag{1}$$

This equation defines $\bar{u} = \rho = \rho\,(r, \theta, \varphi) = \rho\,(r, \theta)$.
Now

$$A_r + A_\rho \frac{\partial\rho}{\partial r} = \frac{2}{r} A, \quad A_\rho \frac{\partial\rho}{\partial\theta} = 2\pi r^2 \sin\theta \;, \quad \frac{\partial\rho}{\partial\varphi} = 0 \quad , \tag{2}$$

so that (cf. C 4 (1))

$$\begin{aligned} \text{grad } \bar{u} &= \frac{2A/r - A_r}{A_\rho} \vec{t} + \frac{2\pi r \sin\theta}{A_\rho} \vec{t}_1 \;, \\ |\text{grad } \bar{u}|^2 &= \frac{(2A/r - A_r)^2 + 4\pi^2 r^2 \sin^2\theta}{A_\rho^2} \quad . \end{aligned} \tag{3}$$

This quantity has to be multiplied by dr and by the area of the zone between θ and $\theta + d\theta$ which is

$$2\pi r \sin\theta \,.\; r\, d\theta = A_\rho \;\, d\rho \quad . \tag{4}$$

Hence we obtain for the element of Dirichlet's integral

$$\frac{(2A/r - A_r)^2 + A(4\pi - A/r^2)}{A_\rho} \, dr \, d\rho \quad . \tag{5}$$

We used here the identity $4\pi^2 r^2 \sin^2\theta = A(4\pi - A/r^2)$.

(b) Comparing the final expression of C 4 and (5) we see that it suffices to prove the following inequalities:

$$\frac{(2A/r - A_r)^2}{A_\rho} \leqq \int_{s(r,\rho)} f^2(g^2 + h^2)^{-1/2} \, ds \quad ,$$

(6)

$$\frac{A(4\pi - A/r^2)}{A_\rho} \leqq \int_{s(r,\rho)} (g^2 + h^2)^{1/2} \, ds \quad .$$

We obtain the first by applying Schwarz's inequality to C 4 (6) and taking C 4 (9) into account.

As to the second inequality we apply Schwarz's inequality again:

$$[L(r,\rho)]^2 = [\int_{s(r,\rho)} ds]^2 \leqq \int_{s(r,\rho)} (g^2 + h^2)^{1/2} ds \int_{s(r,\rho)} (g^2 + h^2)^{-1/2} \, ds \tag{7}$$

where $L(r,\rho)$ is the length of the curve $s(r,\rho)$. But according to the isoperimetric inequality on the sphere*

$$[L(r,\rho)]^2 \geqq A(4\pi - A/r^2) \quad ; \tag{8}$$

combining C 4 (9), (7) and (8), we obtain the second inequality (6).

We observe that the isoperimetric inequality remains valid for a system of separated spherical curves as it is easy to show.

---

* See W. Blaschke, Vorlesungen über Differentialgeometrie, vol. 1, p. 51.

NOTE D. ON A GENERALIZATION OF DIRICHLET'S INTEGRAL

D 1. Several questions in Mathematical Physics lead to the problem to minimize the integral

$$\iint \{|\text{grad } u|^2 + p(x,y)u^2\} \, d\sigma \quad ; \quad d\sigma = dx\, dy \quad . \tag{1}$$

Here the integration is extended over a certain domain $\underline{D}$ in the x,y-plane, p(x,y) is a given function defined in $\underline{D}$ and the admissible functions u assume given values on the boundary of $\underline{D}$. The corresponding Euler-Lagrange differential equation is*

$$\nabla^2 u = p(x,y)\, u \ . \tag{2}$$

We obtain the classical Dirichlet problem if p(x,y) = 0. This special case occurs in the study of the distribution of electricity on a cylindrical condenser. The domain $\underline{D}$ is in this case ring-shaped, that is, it is bounded by two curves $\underline{C}_0$ and $\underline{C}_1$ such that $\underline{C}_0$ is completely in the interior of $\underline{C}_1$. The boundary values are: u = 0 on $\underline{C}_0$ and u = 1 on $\underline{C}_1$. If we denote the minimum of (1) (in the case p = 0) by $4\pi c$, we call c the logarithmic capacity of the plane domain $\underline{D}$ or the capacity per unit length of the cylindrical condenser having $\underline{D}$ as cross-section. Another interpretation of c is related to the thermal conductance of a plate whose shape is that of $\underline{D}$.

D 2. Concerning this capacity constant c we mention the following result of Carleman (2): Of all domains D whose boundary curves C_0 and C_1 have given areas, the capacity c will be a minimum if the curves C_0 and C_1 are concentric circles of the given areas.

This theorem describes the change of c under symmetrization of the boundary curves of $\underline{D}$ with respect to a given point. The theorem is an extension of the area theorems on conformal mapping. The proof of Carleman is based on conformal mapping using the fact that c is a conformal invariant. Other proofs use the idea of symmetrization. The corresponding result for symmetrization with respect to a line was obtained by Pólya-Szegö (24).

---

* See, for instance, S. Bergmann and M. Schiffer, Kernel functions in the theory of partial differential equations of elliptic type, Duke Math. Journal, vol. 15 (1948) pp. 535-566.

D 3. Theorem on the generalized capacity. Before attempting to extend the theorem of Carleman to the more general integral D 1 (1), we make the special assumption that the function $p(x,y)$ has circular symmetry, that is, it depends only on the distance r from a fixed point in the interior of C_0. We can assume that this point is the origin, $p(x,y) = p(r)$. Assuming moreover that $p(r)$ is decreasing for $r > 0$ and that $r\,p(r)$ is integrable in a neighborhood of $r = 0$, we prove the following theorem:

We denote by $4\pi c$ the minimum of the generalized Dirichlet integral

$$(1) \qquad D(u) = \iint_D \{|\text{grad } u|^2 + p(r)\, u^2\}\, d\sigma - \iint_G p(r)\, d\sigma \quad ;$$

here the first integral is extended over the ring-shaped domain D bounded by the inner curve C_0 and the outer curve C_1, the second integral over the whole interior G of C_1 and u satisfies the boundary conditions $u = 0$ on C_0 and $u = 1$ on C_1.

Of all ring-shaped domains D with given area and with given area of the containing simply connected domain G, the annulus contained between two concentric circles has the minimum generalized capacity c.

The quantity c thus defined is a generalization of the logarithmic capacity which corresponds to the case $p(r) = 0$.

D 4. The proof of the theorem formulated in D 3 is based on the process of symmetrization with respect to a point. This means the replacement of a given curve by a circle of the same area with center at the given point. We can assume that the function $p(r)$ is positive. Indeed, if we add a constant k to $p(r)$ the change of the integral $D(u)$ will be

$$k \iint_D u^2 d\sigma \; - k \iint_G d\sigma \; .$$

Both integrals remain unchanged by symmetrization.

Let u be the minimizing function of $D(u)$, $u = 0$ on C_0 and $u = 1$ on C_1. We denote the level sets $u = \rho$ by C_ρ and assume first that they are simple curves of the same distribution as in the harmonic case $\nabla^2 u = 0$; that is, we assume that $0 \leq \rho \leq 1$ and C_ρ is contained in $C_{\rho'}$ provided that $\rho < \rho'$. We denote the area of C_ρ by $A(\rho) = \pi R^2$. Hence $R = R(\rho)$ is the radius of the circle which has the same area as C_ρ.

We "symmetrize" the curves C_ρ and the function u in the following manner. We replace C_ρ by the circle $\bar{C}_\rho$ of radius R about the origin and define $\bar{u}$ by the condition that $\bar{u} = \rho$ on the circle $\bar{C}_\rho$. Obviously $\bar{u}$ is an admissible function of the annulus $\bar{D}$ bounded by the concentric circles $\bar{C}_0$ and $\bar{C}_1$.

We show that this process diminishes the generalized Dirichlet's integral D(u). Thus the same will be true for the minimum of D(u) for all admissible u. This will yield the assertion.

D 5. We prove first that the process described in D 4 diminishes Dirichlet's integral

$$\iint_{\underline{D}} |\text{grad } u|^2 \, d\sigma \quad . \tag{1}$$

(This leads to the proof of the theorem of Carleman.) For the sake of completeness we give here a proof based on the idea of symmetrization; the basic idea of this proof has been applied in many cases (for instance in 3.2). We show that by the process in question not only the integral (1) is diminished, but even an infinitesimal element of it corresponding to the ring-shaped domain between $\underline{C}_\rho$ and $\underline{C}_{\rho+d\rho}$, $d\rho > 0$.

Indeed, we may write the area element corresponding to the points on $\underline{C}_\rho$ in the form $d\sigma = ds.dn = ds.|\text{grad } u|^{-1} d\rho$ where ds is the arc-element of $\underline{C}_\rho$. The contribution of the elementary ring-shaped domain will be

$$d\rho \int_{\underline{C}_\rho} |\text{grad } u| \, ds \quad . \tag{2}$$

This has to be compared with

$$d\rho \int_{\overline{C}_\rho} |\text{grad } \bar{u}| \, ds = d\rho \ . \ |\text{grad } \bar{u}| \ .2\pi R \quad . \tag{3}$$

But $|\text{grad } \bar{u}| = \frac{d\bar{u}}{dR} = \frac{d\rho}{dR}$ so that (3) is identical with

$$d\rho \ . \ \frac{d\rho}{dR} \ . \ 2\pi R = d\rho \ . \ \frac{d\rho}{d(R^2)} \ . \ 4\pi R^2 = d\rho \ . \ \frac{4\pi A(\rho)}{A'(\rho)}$$

Now the area of the ring-shaped domain is

$$A(\rho + d\rho) - A(\rho) = A'(\rho) d\rho$$

or

$$A'(\rho) d\rho = \int_{\underline{C}_\rho} ds.dn = d\rho \int_{\underline{C}_\rho} |\text{grad } u|^{-1} ds \quad . \tag{4}$$

Using the inequality of Schwarz we obtain

$$(5)\qquad \int_{\underline{C}_\rho} |\text{grad } u|\, ds \int_{\underline{C}_\rho} |\text{grad } u|^{-1}\, ds \geqq \left[\int_{\underline{C}_\rho} ds\right]^2 = [L(\rho)]^2$$

where $L(\rho)$ is the length of $\underline{C}_\rho$. According to the isoperimetric inequality $[L(\rho)]^2 \geqq 4\pi A(\rho)$ so that indeed

$$(6)\qquad \int_{\underline{C}_\rho} |\text{grad } u|\, ds \geqq \frac{4\pi A(\rho)}{A'(\rho)} = \int_{\overline{C}_\rho} |\text{grad } \bar{u}|\, ds \quad .$$

D 6. We consider now the second part of $D(u)$:

$$(1)\qquad E(u) - E = \int_{\underline{D}} p(r)\, u^2 d\sigma - \int_{\underline{G}} p(r) d\sigma$$

which involves the function $p(r)$. The first integral $E(u)$ depends on u, the second E not. Under the assumption $p(r) > 0$ made above this difference $E(u) - E$ is negative. We introduce a new function q defined by the condition

$$(2)\qquad q'(\pi r^2) = p(r) \ , \quad q(0) = 0 \quad ,$$

or

$$(3)\qquad q(\pi r^2) = 2\pi \int_0^r p(r) r\, dr \quad .$$

(Here we used the integrability property of $p(r)$ assumed in D 3.) The function $q(t)$ is <u>convex from above</u>, since $q'(t)$ is a decreasing function. We assume first that all curves $\underline{C}_\rho$ (including $\underline{C}_0$ and $\underline{C}_1$) are <u>star-shaped with respect to the origin</u>.

Let r, φ be polar coordinates. In the integral $E(u)$ we introduce the coordinates ρ, φ defined by

$$(4)\qquad u = u(r, \varphi) = \rho \quad ;$$

in the integral E we use the polar coordinates r, φ. Let $r = r(\rho, \varphi)$ be the equation of $\underline{C}_\rho$ in polar coordinates. We define the function

$$(5)\qquad Q(\rho) = \frac{1}{2\pi} \int_{-\pi}^{\pi} q[\pi(r(\rho, \varphi))^2] d\varphi$$

so that

(6) $$Q'(\rho) = \int_{-\pi}^{\pi} q'(\pi r^2) r\, r_\rho \, d\varphi = \int_{-\pi}^{\pi} p(r) r\, r_\rho \, d\varphi \quad .$$

Now, since $r_\rho > 0$,

(7) $$d\sigma = r\, dr\, d\varphi = r\, r_\rho \, d\rho \, d\varphi$$

so that the integrals in (1) can be written as follows:

(8) $$E(u) = \int_{-\pi}^{\pi} \int_{\rho=0}^{1} p(r) \rho^2 \, r\, r_\rho \, d\rho \, d\varphi \ ,$$
$$E = \int_{-\pi}^{\pi} \int_{r=0}^{r(1,\varphi)} p(r)\, r\, dr\, d\varphi \quad .$$

By (3)

(9) $$\int_{r=0}^{r(1,\varphi)} p(r)\, r\, dr = \frac{1}{2\pi} q\, [\pi(r(1,\varphi))^2]$$

so that (8) can be rewritten as follows:

(10) $$E(u) = \int_0^1 \rho^2 Q'(\rho)\, d\rho \ , \quad E = Q(1) \quad .$$

Here we used (5) and (6). Now

(11) $$E(u) - E = \int_0^1 \rho^2 Q'(\rho) d\rho - Q(1) = -2 \int_0^1 \rho\, Q(\rho) d\rho \quad .$$

D 7. The corresponding quantities for the function $\bar{u}$, defined in the symmetrized domain $\bar{D}$, can be easily computed using the equation $r = R = [\pi^{-1} A(\rho)]^{1/2}$ of the circle $\bar{C}_\rho$. We obtain

(1) $$\bar{Q}(\rho) = \frac{1}{2\pi} \int_{-\pi}^{\pi} q(\pi R^2) d\varphi = q(\pi R^2) = q\left[A(\rho)\right] \quad .$$

We apply now Jensen's inequality to the convex function $q(t)$ where $r = r(\rho, \varphi)$:

$$(2) \qquad \bar{Q}(\rho) = q\left[A(\rho)\right] = q\left[\frac{1}{2\pi}\int_{-\pi}^{\pi} \pi r^2 d\varphi\right] \geqq \frac{1}{2\pi}\int_{-\pi}^{\pi} q(\pi r^2)d\varphi = Q(\rho) \quad .$$

Hence

$$-2\int_0^1 \rho\, Q(\rho)d\rho \geqq -2\int_0^1 \rho\, \bar{Q}(\rho)d\rho \ .$$

This establishes the assertion.

We observe that in view of (2) the integral E is increased (not decreased) by symmetrization. (This follows also from the theorem in sect. 7.2.) In the special case when the exterior curve is a circle about the origin, we have $Q(1) = q\left[A(1)\right] = \bar{Q}(1)$ so that E remains unchanged in this case. The first integral E(u) is then decreased.

D 8. The proof above is based on the following assumptions:

(a) The level sets C_ρ are simple curves distributed in the same way as in the harmonic case $\nabla^2 u = 0$.

(b) The level curves C_ρ are star-shaped with respect to the origin.

We have also assumed (and this was an unessential assumption) that $p(r) > 0$.

These restrictions can be removed. For this and further details we refer to Szegö (35).

NOTE E. HEAT-CONDUCTION ON A SURFACE

E 1. Problem. We consider an arbitrary open or closed surface in the three-dimensional Euclidean space represented by the parameters u, v. We use the classical symbols E, F, G for the fundamental quantities of the first kind; we have

$$d\sigma = (EG - F^2)^{1/2}\, du\, dv = W\, du\, dv \tag{1}$$

for the area-element of the surface.

Let $\underline{D}$ be any connected domain on this surface bounded by one or more closed curves. If f is any function defined on the surface, the integral

$$\iint_{\underline{D}} |\text{grad } f|^2\, d\sigma \tag{2}$$

is called Dirichlet's integral over $\underline{D}$. (The gradient is defined in the sense of the metric on the surface, see below.) This quantity is obviously a conformal invariant.

We consider two closed curves on the surface, $\underline{C}_0$ and $\underline{C}_1$ which jointly bound a ring-shaped domain $\underline{D}$. We shall call $\underline{C}_0$ the "inner" curve and $\underline{C}_1$ the "outer" curve. We define the conductance c of this domain in the following manner: $4\pi c$ is the minimum of the Dirichlet integral (2) extended over $\underline{D}$ provided that we admit all (sufficiently smooth) functions f defined on $\underline{D}$ and satisfying the boundary conditions:

$$f = 0 \quad \text{on} \quad \underline{C}_0 \; ; \quad f = 1 \quad \text{on} \quad \underline{C}_1 \quad . \tag{3}$$

Our task is to find lower bounds for the quantity c thus defined in terms of geometrical quantities connected with $\underline{D}$.

The simplest special case is that of a plane domain $\underline{D}$ in which case c is the "logarithmic capacity" (capacity per unit length of a cylindrical condenser; see 2.3, 2.7 A). Another interesting case is that of a spherical domain.*

---

* This case has been considered by E. Zermelo in connection with a study of the dynamics of cyclones. See, Hydrodynamische Untersuchungen über die Wirbelbewegungen in einer Kugelfläche, Zeitschrift für Math. und Physik, vol. 47 (1902) pp. 201-237. - See also D. Hilbert, Grundzüge einer allgemeinen Theorie der linearen Integralgleichungen, 1912, p. 65.

Generally, c is proportional to the thermal or electrical conductance of a curved shell of shape $\underline{D}$. In the electrical interpretation $\underline{C}_0$ and $\underline{C}_1$ are the "electrodes."

E 2. Green's formula. It is known* that

$$W^2 \, |\text{grad } f|^2 = E \, f_v^{\,2} - 2 \, F \, f_v \, f_u + G \, f_u^{\,2} \tag{1}$$

so that introducing the standard notation

$$D(f,g) = \iint_{\underline{D}} (E \, f_v \, g_v - F \, (f_v \, g_u + f_u \, g_v) + G \, f_u \, g_u) W^{-2} \, d\sigma \tag{2}$$

we can write the Dirichlet integral E 1 (2) in the simple form D(f,f). Introducing the Laplacian

$$\nabla^2 f = \frac{1}{W} \left\{ \left(\frac{E \, f_v - F \, f_u}{W} \right)_v + \left(\frac{G \, f_u - F \, f_v}{W} \right)_u \right\} \tag{3}$$

we have Green's formula

$$D(f,g) = - \int g \, \frac{\partial f}{\partial n_1} \, ds - \iint g \nabla^2 f \, d\sigma \quad . \tag{4}$$

The surface integrals are extended over the given domain and the line integral along its boundary; ds is the arc-element and n_1 is the interior normal. We note two simple consequences of (4):

(a) if $\nabla^2 f = 0$, then

$$\int \frac{\partial f}{\partial n_1} \, ds = 0 \quad ; \tag{5}$$

(b) if $\nabla^2 f = 0$, then

$$D(f,f) = \iint |\text{grad } f|^2 \, d\sigma = - \int f \, \frac{\partial f}{\partial n_1} \, ds \quad . \tag{6}$$

Let now f be the minimizing function of Dirichlet's integral E 1 (2) under the conditions E 1 (3). We have by (6) and (5):

$$4\pi c = - \int \frac{\partial f}{\partial n} \, ds \quad . \tag{7}$$

---

* See for instance, W. Blaschke, Vorlesungen über Differentialgeometrie, vol. 1, § § 66-68, pp. 112-116.

The integration is extended over an arbitrary curve which divides the ring-shaped domain between $\underline{C}_0$ and $\underline{C}_1$ into two ring-shaped domains and can be deformed into $\underline{C}_0$ or into $\underline{C}_1$; the normal is directed from $\underline{C}_1$ to $\underline{C}_0$. Another way of writing this is the following:

$$(8) \qquad 4\pi c = \int |\text{grad } f| \, ds$$

where the integration is extended over any level curve of f in $\underline{D}$.

If E = G, F = 0, the mapping of the surface onto the u, v-plane is conformal. In this case

$$W = E, \quad d\sigma = E \, du \, dv, \quad \nabla^2 f = E^{-1} (f_{uu} + f_{vv})$$

by (3), and

$$D(f,f) = \iint_{\underline{D}} (f_u^{\,2} + f_v^{\,2}) \, du \, dv$$

by (2). Therefore, c <u>remains invariant under conformal mapping</u>.

E 3. <u>Lower bound for</u> c. Concerning the following argument cf. sect. 3.2 and notes C and D.

We consider the level curves $\underline{C}_\rho$ of the minimizing function f defined by $f = \rho$, $0 \leqq \rho \leqq 1$. We denote by $L(\rho)$ the length of $\underline{C}_\rho$ and by $A(\rho)$ the area between $\underline{C}_0$ and $\underline{C}_\rho$. The area of the ring-shaped domain between $\underline{C}_\rho$ and $\underline{C}_{\rho + d\rho}$, $d\rho > 0$, is

$$(1) \qquad A(\rho + d\rho) - A(\rho) = A'(\rho) d\rho = \int ds.dn$$

where the integration is extended over $\underline{C}_\rho$ and dn denotes the piece of the normal of $\underline{C}_\rho$ between $\underline{C}_\rho$ and $\underline{C}_{\rho + d\rho}$. Since $|\text{grad } f| = d\rho/dn$ we find

$$(2) \qquad A'(\rho) = \int_{\underline{C}_\rho} |\text{grad } f|^{-1} \, ds \quad .$$

By Schwarz's inequality:

$$(3) \qquad 4\pi c \, A'(\rho) = \int_{\underline{C}_\rho} |\text{grad } f| \, ds \int_{\underline{C}_\rho} |\text{grad } f|^{-1} \, ds \geqq \left[\int_{\underline{C}_\rho} ds \right]^2 = \left[L(\rho) \right]^2 .$$

From this inequality lower bounds for c can be obtained under various special assumptions. We shall discuss the cases of a plane domain, a spherical domain and a cylindrical domain. These domains can be mapped conformally onto each other, and such a mapping does not affect c (sect. E 2). This yields inequalities for the logarithmic capacity which are different from that of Carleman (2).

E 4. Plane domain. Let P_0 and P_1 be the boundary curves of a plane domain D, P_0 is contained in the interior of P_1. It is convenient to assume that P_0 contains the origin. Denoting by c the conductance of D we have

$$c^{-1} \leqq \log |P_1| - \log |P_0| \tag{1}$$

where $|P_0|$ and $|P_1|$ denote the areas of the domains bounded by P_0 and P_1, respectively.

This inequality is due to Carleman (2). It follows readily from E 3 (3) by applying the isoperimetric inequality $[L(\rho)]^2 \geqq 4\pi A(\rho)$. Indeed, we obtain

$$4\pi c\, A'(\rho) \geqq 4\pi\, A(\rho) \;, \quad c^{-1} \leqq \frac{A'(\rho)}{A(\rho)} \;. \tag{2}$$

Now we integrate with respect to ρ, $0 \leqq \rho \leqq 1$.

In case of an annular domain bounded by concentric circles with radii r_0 and r_1, $r_0 < r_1$, we have

$$c^{-1} = \log (\pi r_1^2) - \log (\pi r_0^2) = 2 \log \frac{r_1}{r_0} \tag{3}$$

so that in this case the sign = holds in (1).

E 5. Spherical domain. Let D be a domain on the unit sphere bounded by the spherical curves S_0 and S_1. We assume that the pole is separated by S_1 from S_0. We interpret the area of any closed spherical curve S not passing through the pole as the area of that spherical domain bounded by S which does not contain the pole; we denote this area by $|S|$.

(a) For the conductance c of the spherical domain D bounded by S_0 and S_1 we have the inequality

$$c^{-1} \leqq \log \frac{|S_1|}{4\pi - |S_1|} - \log \frac{|S_0|}{4\pi - |S_0|} \;. \tag{1}$$

This inequality becomes an equation if S_0 and S_1 are concentric circles.

We derive inequality (1) from E 3 (3) by using the isoperimetric inequality*

$$[L(\rho)]^2 \geqq A(\rho)\ [\, 4\pi - A(\rho)\,] \;. \tag{2}$$

We obtain

---

* W. Blaschke, loc. cit., p. 51.

(3) $$c^{-1} \leqq \frac{A'(\rho)}{A(\rho)} + \frac{A'(\rho)}{4\pi - A(\rho)}$$

so that by integration

(4) $$c^{-1} \leqq \left[\log \frac{A(\rho)}{4\pi - A(\rho)}\right]_0^1$$

follows.

(b) We choose as usual the parameters $u = \theta$, $v = \varphi$ on the unit sphere where θ is the spherical distance from the pole and φ the meridian angle; we have then $E = 1$, $F = 0$, $G = \sin^2 \theta$, $W = \sin \theta$ and

(5) $$\sin \theta \ \nabla^2 f = (\sin \theta)^{-1} f_{\varphi\varphi} + (\sin \theta \ f_\theta)_\theta \quad .$$

Let $f = f(\theta)$ be a function of θ alone. Then $\nabla^2 f = 0$ can be integrated and we obtain $f(\theta) = A \log \operatorname{ctg}(\theta/2) + B$. Let $\theta = \theta_0$ and $\theta = \theta_1$ be the circles $\underline{S}_0$ and $\underline{S}_1$, $\theta_0 > \theta_1$. We have, choosing $A = 1$, $B = 0$,

$$4\pi c \left(f(\theta_1) - f(\theta_0)\right) = - \int \frac{\partial f}{\partial n} ds = - \int \frac{\partial f}{\partial \theta} \sin \theta \, d\varphi = 2\pi$$

so that

(6) $$c^{-1} = \log \operatorname{ctg}^2 \frac{\theta_1}{2} - \log \operatorname{ctg}^2 \frac{\theta_0}{2} \quad .$$

On the other hand,

(7) $$|\underline{S}_0| = \int_{\theta_0}^{\pi} \int_{-\pi}^{\pi} \sin \theta \, d\theta \, d\varphi = 2\pi(1 + \cos \theta_0) = 4\pi \cos^2 \frac{\theta_0}{2}$$

and a similar formula holds for $|\underline{S}_1|$. This shows that in this particular case equality is attained in (1).

E 6. <u>Cylindrical domain</u>. Another interesting case is that of a right circular cylinder. We assume that the radius of any orthogonal cross-section is 1, and consider a domain $\underline{D}$ on the cylinder bounded by the cylindrical curves $\underline{C}_0$ and $\underline{C}_1$. All such curves considered henceforth are assumed to be deformable by continuous transition into an orthogonal cross-section. The curves $\underline{C}_0$ and $\underline{C}_1$ shall not have any points in common.

<u>For the conductance c of the cylindrical domain D bounded by the curves</u> $\underline{C}_0$ and $\underline{C}_1$ <u>we have the inequality</u>

(1) $$c^{-1} \leqq \pi^{-1} \, |\underline{C}_0, \underline{C}_1|$$

where $|C_0, C_1|$ is the area of the domain D.

The sign = holds if C_0 and C_1 are orthogonal cross-sections.

This follows again from E 3 (3) since the length of any level curve is $\geq 2\pi$. Thus

(2) $$4\pi c A'(\rho) \geq (2\pi)^2$$

and (1) is obtained by integration.

For the discussion of the sign = we introduce conveniently the angle v in a fixed orthogonal cross-section and the distance u measured along the generators from that cross-section, as parameters: $-\infty < u < \infty$, $-\pi < v \leq \pi$. We have $E = G = 1$, $F = 0$ so that $\nabla^2 f$ is the ordinary Laplacian. Hence $f = Au + B$ is harmonic. If u_0 and u_1 are the coordinates of the circles C_0 and C_1 which are parallel cross-sections, $u_0 < u_1$, we have $2c = (u_1 - u_0)^{-1}$ and the area of D is $2\pi(u_1 - u_0)$. This confirms the assertion regarding the sign = in (1).

Another way of writing (1) is the following:

(3) $$c \geq \frac{\ell^2}{4\pi A}$$

where A is the area of the given cylindrical domain D and ℓ is the perimeter of the orthogonal cross-section. (Indeed, $\ell = 2\pi$.) In this form the result is independent of the assumption that the radius of the orthogonal cross-section is 1.

E 7. *Application to the conductance of a plane domain.* The conductance is a conformal invariant (E 2) so that the results obtained for the sphere and cylinder can be applied to the case of plane domains.

We consider again a plane domain D bounded by the curves P_0 and P_1 where, as in E 4, P_0 is contained in P_1 and P_0 contains the origin. We project the plane stereographically onto the unit sphere (see E 8). Denoting the resulting curves by S_0 and S_1, the mutual position of these curves and their relative position to the pole will be the same as outlined in E 5.

Similarly we can draw the Mercator projection of D onto a cylinder with the unit circle as orthogonal cross-section (see E 8). The images C_0 and C_1 of P_0 and P_1 will satisfy the conditions of E 6.

Using the previous notation and taking into account the conformal invariance of the capacity, we obtain for the conductance c of the plane domain D the inequalities:

(1) $$c^{-1} \leq \log |P_1| - \log |P_0| \quad ,$$

(2) $$c^{-1} \leqq \log \frac{|\underline{S}_1|}{4\pi - |\underline{S}_1|} - \log \frac{|\underline{S}_0|}{4\pi - |\underline{S}_0|} ,$$

(3) $$c^{-1} \leqq \pi^{-1} |\underline{C}_1, \underline{C}_0| .$$

The first inequality is due to Carleman (2), the second was proved in E 5, the third in E 6.

E 8. Analytic formulation. The natural question arises which of the inequalities proved in E 7 is more informative. In order to facilitate this comparison, we make use of the well-known analytic form of the stereographic and Mercator projection.*

(a) Let $x + iy = re^{i\varphi}$ be the complex plane in which $\underline{D}$ is considered. We denote by ξ, η, ζ the space in which the unit sphere is situated; let θ, φ $(0 \leqq \theta \leqq \pi, -\pi < \varphi \leqq \pi)$ be the spherical coordinates on this sphere. (The angle φ does not change under this projection so that the same symbol can be used in the plane and on the sphere.) We have then for the stereographic projection:

(1) $$x = \frac{\xi}{1-\zeta} = \operatorname{ctg}\frac{\theta}{2}\cos\varphi , \quad y = \frac{\eta}{1-\zeta} = \operatorname{ctg}\frac{\theta}{2}\sin\varphi ,$$
$$dx\,dy = (1 - \cos\theta)^{-2}\,d\sigma ,$$

where $d\sigma$ is the area-element on the sphere. If $\underline{P}$ is an arbitrary closed curve in the plane containing the origin and $\underline{S}$ the image of $\underline{P}$ on the sphere, we have

(2) $$|\underline{S}| = \iint d\sigma = \iint (1 - \cos\theta)^2\,dx\,dy$$
$$= \iint 4(1 + x^2 + y^2)^{-2}\,dx\,dy = \iint 4(1 + r^2)^{-2}\,r dr\,d\varphi$$

the latter integration being extended over the interior of $\underline{P}$ (cf. the convention laid down in E 5).

It is sometimes convenient to represent the curve $\underline{P}$ in polar coordinates $r = R = R(\varphi)$. In all cases when such a representation is used, we assume that the curve in question is star-shaped with respect to the origin. We find

* See, for instance, G. Pólya-G. Szegö, Aufgaben und Lehrsätze, vol. 1 (1925) p. 95, problems 60 and 63; see also pp. 266-267.

$$|\underline{S}| = \int_{-\pi}^{\pi} \int_{0}^{R(\varphi)} 4(1+r^2)^{-2}\, r\, dr\, d\varphi = 2 \int_{-\pi}^{\pi} [R(\varphi)]^2 \left\{1+[R(\varphi)]^2\right\}^{-1} d\varphi ,$$

(3)

$$\frac{|\underline{S}|}{4\pi - |\underline{S}|} = \int_{-\pi}^{\pi} [R(\varphi)]^2 \left\{1+[R(\varphi)]^2\right\}^{-1} d\varphi \ : \ \int_{-\pi}^{\pi} \left\{1+[R(\varphi)]^2\right\}^{-1} d\varphi .$$

(b) In order to deal with the Mercator projection we spread out the cylinder into a strip of the complex $u + iv$-plane. The Mercator projection is then equivalent with the conformal mapping

(4) $$x + iy = e^{u+iv}$$

If $\underline{P}_0$, $\underline{P}_1$, $\underline{C}_0$, $\underline{C}_1$ have the previous meaning, we find

(5) $$|\underline{C}_0, \underline{C}_1| = \iint du\, dv = \iint \frac{dx\, dy}{x^2 + y^2} = \iint \frac{dr\, d\varphi}{r} ,$$

the latter integration being extended over the domain between $\underline{P}_0$ and $\underline{P}_1$.

Representing the curves $\underline{P}_0$ and $\underline{P}_1$ in polar coordinates $r = R_0(\varphi)$ and $r = R_1(\varphi)$, $0 < R_0(\varphi) < R_1(\varphi)$, we find

(6) $$|\underline{C}_0, \underline{C}_1| = \int_{-\pi}^{\pi} \log \frac{R_1(\varphi)}{R_0(\varphi)}\, d\varphi .$$

Thus we find the following alternate forms of the inequalities E 7 (1), (2), (3):

(7) $$\exp(c^{-1}) \leq \int_{-\pi}^{\pi} [R_1(\varphi)]^2\, d\varphi \ : \ \int_{-\pi}^{\pi} [R_0(\varphi)]^2\, d\varphi ,$$

(8) $$\exp(c^{-1}) \leq \frac{\int_{-\pi}^{\pi} [R_1(\varphi)]^2\left\{1+[R_1(\varphi)]^2\right\}^{-1} d\varphi}{\int_{-\pi}^{\pi} \left\{1+[R_1(\varphi)]^2\right\}^{-1} d\varphi} \ : \ \frac{\int_{-\pi}^{\pi} [R_0(\varphi)]^2\left\{1+[R_0(\varphi)]^2\right\}^{-1} d\varphi}{\int_{-\pi}^{\pi} \left\{1+[R_0(\varphi)]^2\right\}^{-1} d\varphi} ,$$

(9) $$\exp(c^{-1}) \leq \exp\left\{\frac{1}{\pi} \int_{-\pi}^{\pi} \log \frac{R_1(\varphi)}{R_0(\varphi)}\, d\varphi\right\} .$$

A further inequality can be obtained by applying to $\underline{D}$ a transformation by reciprocal radii. This transformation replaces R_0 and R_1 by R_1^{-1} and

and R_0^{-1}, respectively, so that from (7)

(10) $$\exp(c^{-1}) \leq \int_{-\pi}^{\pi} [R_0(\varphi)]^{-2}\, d\varphi \ : \ \int_{-\pi}^{\pi} [R_1(\varphi)]^{-2}\, d\varphi$$

results. The same transformation does not change the inequalities (8) and (9).

E 9. <u>Inner and outer radius</u>. These radii have been defined in 2.3 (2) and (3) as limiting cases of the capacity (conductance). The inequalities of E 7 and E 8 can be used immediately, resulting in the following information.

Let $\underline{P}$ be an arbitrary curve in the plane surrounding the origin. We denote by $\underline{S}$ the stereographic image of $\underline{P}$ on the sphere and by $\underline{C}$ the Mercator image of $\underline{P}$ on the cylinder. Using the same notation as before we find then for the inner radius r_0 of $\underline{P}$ with respect to the origin and for the outer radius $\bar{r}$ of $\underline{P}$, the following inequalities:

$$\pi r_0^2 \leq |\underline{P}| \leq \pi \bar{r}^2 ,$$

(1) $$\pi r_0^2 \leq \pi \frac{|\underline{S}|}{4\pi - |\underline{S}|} \leq \pi \bar{r}^2$$

$$\pi r_0^2 \leq \pi \exp\left\{\pi^{-1} \, |\underline{C}, \gamma|\right\} \leq \pi \bar{r}^2 .$$

Here $|\underline{C}, \gamma|$ is the "area" of the cylindrical domain between $\underline{C}$ and the intersection γ of the cylinder with the x,y-plane. The elements of this domain have to be taken with the positive or negative sign according as u is positive or negative.

Let now $r = R(\varphi)$ be the equation of $\underline{P}$ in polar coordinates. We find then:

(2) $$r_0 \leq \left\{\frac{1}{2\pi} \int_{-\pi}^{\pi} [R(\varphi)]^2\, d\varphi\right\}^{1/2} \leq \bar{r} ,$$

(3) $$r_0 \leq \left\{\frac{\int_{-\pi}^{\pi} [R(\varphi)]^2 \left\{1 + [R(\varphi)]^2\right\}^{-1} d\varphi}{\int_{-\pi}^{\pi} \left\{1 + [R(\varphi)]^2\right\}^{-1} d\varphi}\right\}^{1/2} \leq \bar{r} ,$$

(4) $$r_0 \leq \exp\left\{\frac{1}{2\pi} \int_{-\pi}^{\pi} \log R(\varphi)\, d\varphi\right\} \leq \bar{r} ,$$

$$(5) \qquad r_0 \leq \left\{ \frac{1}{2\pi} \int_{-\pi}^{\pi} [R(\varphi)]^{-2}\, d\varphi \right\}^{-1/2} \leq \bar{r} .$$

E 10. Comparison. It is interesting to compare the previous inequalities with each other. Let us start with the inequalities of E 8.

(a) Inequalities E 8 (7) - (10) are not comparable; i.e., for a suitable choice of the functions $R_0(\varphi)$ and $R_1(\varphi)$, $0 < R_0(\varphi) < R_1(\varphi)$, any of the quantities on the right-hand side can be made larger than any other.

Indeed, taking $R_0(\varphi) = 1$ we find

$$\frac{1}{2\pi} \int_{-\pi}^{\pi} [R_1(\varphi)]^2\, d\varphi \quad \text{and} \quad \left\{ \frac{1}{2\pi} \int_{-\pi}^{\pi} [R_1(\varphi)]^{-2}\, d\varphi \right\}^{-1}$$

for the right-hand expressions in E 8 (7) and (10). Thus the first expression exceeds the second one. Choosing $R_1(\varphi) = 1$ we face the opposite situation.

As to E 8 (8), we replace $R_0(\varphi)$ and $R_1(\varphi)$ by $k^{-1}R_0(\varphi)$ and $k^{-1}R_1(\varphi)$, respectively, where $k > 0$ is a constant. The resulting quantity

$$\frac{\int_{-\pi}^{\pi} [R_1(\varphi)]^2 \left\{k^2 + [R_1(\varphi)]^2\right\}^{-1} d\varphi}{\int_{-\pi}^{\pi} \left\{k^2 + [R_1(\varphi)]^2\right\}^{-1} d\varphi} : \frac{\int_{-\pi}^{\pi} [R_0(\varphi)]^2 \left\{k^2 + [R_0(\varphi)]^2\right\}^{-1} d\varphi}{\int_{-\pi}^{\pi} \left\{k^2 + [R_0(\varphi)]^2\right\}^{-1} d\varphi}$$

yields that on the right-hand side of E 8 (7) as $k \to \infty$ and that on the right-hand side of E 8 (10) as $k \to 0$. This shows that E 8 (8) is not comparable with either E 8 (7) or (10).

As to E 8 (9), it is sufficient to compare it with E 8 (7) and (10). Replacing $R_0(\varphi)$ by

$$\left(\frac{1 - r^2}{1 - 2r\cos\varphi + r^2}\right)^{1/2} R_0(\varphi)$$

and replacing $R_1(\varphi)$ by the corresponding quantity, $0 < r < 1$, we have

$$\lim_{r \to 1} \int_{-\pi}^{\pi} \frac{1 - r^2}{1 - 2r\cos\varphi + r^2} [R_1(\varphi)]^2 d\varphi : \int_{-\pi}^{\pi} \frac{1 - r^2}{1 - 2r\cos\varphi + r^2} [R_0(\varphi)]^2 d\varphi$$

$$= [R_1(0)/R_0(0)]^2$$

But by a proper choice of $R_0(\varphi)$ and $R_1(\varphi)$ this can be made larger or smaller than the mean value on the right of E 8 (9).

(b) We come now to the inequalities of E 9, denoting the middle expressions in E 9 (2) - (5) by A, B, C, D, respectively. Here A is the "square-mean" and C the geometric mean of the function $R(\varphi)$. We write more generally

$$(1) \qquad M_p = \left\{ \frac{1}{2\pi} \int_{-\pi}^{\pi} [R(\varphi)]^p \, d\varphi \right\}^{1/p}$$

so that $A = M_2$, $C = M_0$, $D = M_{-2}$. Hence $A \geqq C \geqq D$. We shall prove that

$$(2) \qquad A \geqq B \geqq D .$$

Also we shall see that B and C are <u>not comparable</u> in general.

The first inequality in (2) is equivalent with

$$(3) \qquad \int_{-\pi}^{\pi} [R(\varphi)]^2 \, d\varphi \int_{-\pi}^{\pi} \left\{ 1 + [R(\varphi)]^2 \right\}^{-1} d\varphi$$

$$\geqq 2\pi \int_{-\pi}^{\pi} [R(\varphi)]^2 \left\{ 1 + [R(\varphi)]^2 \right\}^{-1} d\varphi .$$

This is a special case of Tchebychev's inequality in the extension given by Hardy-Littlewood-Pólya.* The second inequality in (2) follows by replacing $R(\varphi)$ by $[R(\varphi)]^{-1}$.

On the other hand, we define $R(\varphi) = a$ in $-\pi$, 0 and $R(\varphi) = b$ in $0, \pi$ where a and b are two positive constants. Then

$$B^2 = \frac{a^2 (1 + a^2)^{-1} + b^2 (1 + b^2)^{-1}}{(1 + a^2)^{-1} + (1 + b^2)^{-1}} , \qquad C^2 = ab ,$$

$$B^2 - C^2 = \frac{(a - b)^2 (1 - ab)}{2 + a^2 + b^2}$$

so that $B > C$ if $ab < 1$ and $B < C$ if $ab > 1$, $a \neq b$.

Recapitulating, we see that E 9 (2) is equivalent to the classical area theorems and E 9 (5) follows from these through transformation by reciprocal radii. Moreover E 9 (3) is a consequence of $A \geqq B \geqq D$ and E 9 (4) follows from the simple fact that M_p is an increasing function of p. We have, of course, more generally,

$$(4) \qquad r_0 \leqq M_p \leqq \bar{r} , \quad -2 \leqq p \leqq 2 .$$

---

* G. H. Hardy, J. E. Littlewood and G. Pólya, Inequalities, 1934, p. 168.

(c) As an illustration we note that for an ellipse with semi-axes a and b we have, by the area theorem,

$$r_o \leqq (ab)^{1/2} , \tag{5}$$

the expression on the right-hand side being the geometric mean of a and b. Now in this case

$$R(\varphi) = (a^{-2} \cos^2\varphi + b^{-2} \sin^2\varphi)^{-1/2} \tag{6}$$

so that

$$M_{-2} = \left\{ \frac{1}{2\pi} \int_{-\pi}^{\pi} (a^{-2} \cos^2\varphi + b^{-2} \sin^2\varphi) d\varphi \right\}^{-1/2} = \left(\frac{a^{-2} + b^{-2}}{2} \right)^{-1/2} \tag{7}$$

This coincides with the M_{-2}-mean of the numbers a and b. Hence

$$r_o \leqq \left(\frac{a^{-2} + b^{-2}}{2} \right)^{-1/2} = \left(\frac{2\, ab}{a^2 + b^2} \right)^{1/2} \cdot (ab)^{1/2} \tag{8}$$

which is sharper than (5). We note that in this case $\bar{r} = (a + b)/2$.

(d) We can easily ascertain that $p = -2$ and $p = 2$ are the extreme values of p for which (4) generally holds. In other words, if $p > 2$ we can find domains for which $\bar{r} - M_p < 0$, and for $p < -2$ some domains for which $M_p - r_o < 0$ holds. Such examples are furnished by the nearly circular domains considered in 1.32 - 1.35. Their equation is

$$r \leqq R(\varphi) = 1 + \rho(\varphi) = 1 + a_o + 2 \sum_{n=1}^{\infty} (a_n \cos n\varphi + b_n \sin n\varphi) \tag{9}$$

where the coefficients of $\rho(\varphi)$ are infinitesimal quantities of the first order. According to the table in 1.33 we have then*, neglecting terms higher than the second order,

$$\begin{aligned} r_o &= 1 + a_o - \sum_{n=1}^{\infty} (2n + 1)(a_n^2 + b_n^2) , \\ \bar{r} &= 1 + a_o + \sum_{n=1}^{\infty} (2n - 1)(a_n^2 + b_n^2) . \end{aligned} \tag{10}$$

Now

$$[R(\varphi)]^p = 1 + p\rho(\varphi) + \frac{p(p - 1)}{2} [\rho(\varphi)]^2$$

---

* In this table only the interior radius r_c, taken with respect to the centroid of the nearly circular domain, is given. Then the coefficient of $a_1^2 + b_1^2$ is not -3 but +1. The inner radius r_o with respect to the origin is easily obtained (easier than r_c). See 6.3 (18).

so that

$$\frac{1}{2\pi}\int_{-\pi}^{\pi}[R(\varphi)]^p\, d\varphi = 1 + pa_0 + \frac{p(p-1)}{2}\left\{a_0^2 + 2\sum_{n=1}^{\infty}(a_n^2 + b_n^2)\right\} ,$$

$$M_p = 1 + a_0 + (p - 1)\sum_{n=1}^{\infty}(a_n^2 + b_n^2) ,$$

$$\bar{r} - M_p = \sum_{n=1}^{\infty}(2n - p)(a_n^2 + b_n^2) ,$$

$$M_p - r_0 = \sum_{n=1}^{\infty}(2n + p)(a_n^2 + b_n^2) .$$

This leads to the conclusion formulated above.

For instance: $\bar{r} - M_p < 0$ holds for the nearly circular domain $r \leqq 1 + \delta \cos\varphi$ provided $p > 2$ and δ is sufficiently small. For the same domain $M_p - r_0 < 0$ holds provided $p < -2$.

The latter domain differs of course from a circle only in terms of second order in δ , and both assertions are trivial for a circle containing the origin but with a center different from the origin. Indeed, $\bar{r}$ coincides with the radius of this circle and so does M_2, as it can be shown, no matter where the center of the circle is located. Hence $\bar{r} = M_2$. Since M_p is increasing with p, the difference $\bar{r} - M_p$ must be negative for $p > 2$. On the other hand by inversion of the given circle with respect to the origin (which is different from the center) a new circle arises whose radius is $1/r_0$; the reciprocal of the mean-value M_{-2} of the original circle coincides with the mean-value M_2 of the new circle so that $r_0 = M_{-2}$. Consequently $M_p - r_0$ must be negative for $p < -2$.

NOTE F. ON MEMBRANES AND PLATES

F 1. Definitions. We have discussed the principal frequency Λ of a membrane with fixed boundary in 1.27 and more thoroughly in ch. 5. We have mentioned briefly the three quantities Λ_1, Λ_2 and Λ_3 ($\Lambda_1 = \Lambda$) in 1.33, and studied their variation in ch. 6. We take up again the study of these quantities.

Let $\underline{D}$ be an arbitrary domain in the plane bounded by a simple analytic curve $\underline{C}$. We denote the area-element of $\underline{D}$ by $d\sigma$. We define three problems of the Calculus of Variations, the admissible function u being defined in $\underline{D}$ and satisfying certain boundary conditions on $\underline{C}$. We write (all the double integrals are taken over $\underline{D}$)

$$\Lambda_1^2 = \min \frac{\iint |\operatorname{grad} u|^2 \, d\sigma}{\iint u^2 \, d\sigma}, \quad u = 0 \text{ on } \underline{C}, \tag{1}$$

$$\Lambda_2^4 = \min \frac{\iint (\nabla^2 u)^2 \, d\sigma}{\iint u^2 \, d\sigma}, \quad u = \frac{\partial u}{\partial n} = 0 \text{ on } \underline{C}, \tag{2}$$

$$\Lambda_3^2 = \min \frac{\iint (\nabla^2 u)^2 \, d\sigma}{\iint |\operatorname{grad} u|^2 \, d\sigma}, \quad u = \frac{\partial u}{\partial n} = 0 \text{ on } \underline{C}. \tag{3}$$

Obviously $\Lambda_2^2 \geq \Lambda_1 \Lambda_3$.

In (1) we allow only continuous functions which have piece-wise continuous first derivatives. In (2) and (3) we allow only functions with continuous first derivatives which have piece-wise continuous second derivatives.

Problems (1) and (2) are classical; the quantities Λ_1 and Λ_2 are the principal frequencies of a membrane with fixed boundary and of a clamped plate, respectively. Problem (3) occurs in the study of the buckling of plates. All three problems have a considerable literature for which we refer to the book of Weinstein (38).

Concerning (1) Lord Rayleigh (26, vol. 1, pp. 339-340) has formulated the following conjecture which was first proved by G. Faber (5) and E. Krahn (14). *Of all membranes of a given area the circle has the gravest fundamental tone (lowest principal frequency).* Lord Rayleigh has formulated the analogous conjecture also for (2); see 26, vol 1, p. 382.

Our purpose is to deal with the analogous problem for (2) and (3), i.e., for the quantities Λ_2 and Λ_3. This can be done under a certain hypothesis concerning the first eigen-functions, that is, for the functions u for which the minima in (2) and (3) are attained. This hypothesis is the following: *The functions u minimizing problems (2) and (3) are different from zero throughout the domain D, or in other words, the first eigen-functions do not have any nodal lines.*

F 2. *Results.* We note the Euler-Lagrange differential equations associated with the minimum problems formulated above:

$$\nabla^2 u + \Lambda_1^2 u = 0, \tag{1}$$

$$\nabla^4 u - \Lambda_2^4 u = 0, \tag{2}$$

$$\nabla^4 u + \Lambda_3^2 \nabla^2 u = 0. \tag{3}$$

It is of interest to point out the minima in question for the special case of a circle. Denoting the radius of the circle by a we have

$$\Lambda_1 = j/a, \quad \Lambda_2 = h/a, \quad \Lambda_3 = k/a \tag{4}$$

where j, h, k denote the smallest positive root of the Bessel functions

$$J_0(x), \quad J_0(x)\, I_0'(x) - J_0'(x)\, I_0(x), \quad J_0'(x), \tag{5}$$

respectively. We have

$$j = 2.4048, \quad h = 3.1962, \quad k = 3.8317. \tag{6}$$

Thus the principal result can be expressed in the following form: *Let D be an arbitrary domain and a the radius of the circle of the same area as D. Then, under the hypothesis formulated above, we have the bounds*

$$\Lambda_1 \geq j/a, \quad \Lambda_2 \geq h/a, \quad \Lambda_3 \geq k/a, \tag{7}$$

or

$$\Lambda_1 \geq j(A/\pi)^{-1/2}, \quad \Lambda_2 \geq h(A/\pi)^{-1/2}, \quad \Lambda_3 \geq k(A/\pi)^{-1/2} \tag{8}$$

where A is the area of D.

The first inequality is the content of the theorem of Rayleigh-Faber-

Krahn, the second and third are proved in the following. In all three cases the proof will be based on the process of symmetrization (with respect to a point) as was essentially the proof of Faber and Krahn.

F 3. A hypothesis. The hypothesis formulated above according to which the first eigen-function of the clamped plate is of constant sign, would follow easily from an affirmative answer to a question of Hadamard raised in a famous investigation (7). This is the question whether or not the Green function $\Gamma(p,q)$ of the differential equation of the clamped plate may change its sign if the points p and q are arbitrary in the interior of the given domain D.* Hadamard himself showed, by using variational methods, that $\Gamma(p,p) > 0$. A discussion of the limiting case of an infinite strip given lately by R. J. Duffin** strongly indicated that the answer to the question of Hadamard is negative in general. Indeed, Ch. Loewner and G. Szegö exhibited recently various cases in which the Green function changes its sign. In their examples the domain D is finite and bounded by a regular analytic curve. Moreoever in these cases Green's function can be computed and discussed in explicit terms.*** A further progress was made by the interesting result of P. R. Garabedian that even the Green function of an ellipse is changing its sign provided the ellipse is sufficiently "flat", that is, its numerical eccentricity comes sufficiently close to 1.****

F 4. Membrane with fixed boundary. The process applied by Faber and Krahn to the case of the membrane, needs essential modification in order to be applicable to the case of the plate. The present section discusses the problem of the membrane with this modified method.

(a) We denote by u the minimizing function of the problem F 1 (1) which is known to be different from zero in the interior of the domain D. Hence we can assume that $0 \leq u \leq 1$. We denote the level set $u = \rho$ by C_ρ, $0 \leq \rho \leq 1$, so that $C_0 = C$ and C_1 coincides with the point (points) at which the maximum $u = 1$ is attained. The set C_ρ consists in general of a finite number of separated curves, and we have $u > \rho$ in the interior of these curves. For the sake of simplicity we assume that C_ρ consists of a single curve. (This restriction will be removed in sect. F 5.)

---

* As to the definition of this Green function see, for instance, Ph. Frank-R. v. Mises, Die Differential-und Integralgleichungen der Mechanik und Physik, vol. 1, 2d. edition, 1930, pp. 856-862.

** R. J. Duffin, On a question of Hadamard concerning super-biharmonic functions, Journal of Math. and Physics, vol. 27 (1949) pp. 253-258.

*** Ch. Loewner, On generations of solutions of the biharmonic equation in the plane by conformal mappings. With two notes by G. Szegö. Technical Report No. 13, Office of Naval Research, Stanford University, August 26, 1950.

**** P. R. Garabedian, A partial differential equation arising in conformal mapping. Technical Report No. 8, Office of Naval Research, Stanford University, August 5, 1950.

Let us denote by $A(\rho)$ the area of the domain inside of $\underline{C}_\rho$. Thus $A(0) = A$ is the area of the whole domain $\underline{D}$ and $A(1) = 0$. We define the radius R by the equation $A(\rho) = \pi R^2$ so that R is a decreasing function of ρ; the maximum R_0 of R is the radius of the circle $\bar{D}$ which has the same area as the given domain $\underline{D}$.

We symmetrize the level curves $\underline{C}_\rho$ by replacing $\underline{C}_\rho$ by a circle of radius R about a fixed point, say the origin. The domain $\underline{D}$ is replaced then by the circular disk $\bar{D}$ of radius R_0. We define on $\bar{D}$ a function $\bar{u}$ by the condition that $\bar{u} = f(\rho)$ on the circle of radius R; the function f will be determined in such a way that the integral in the numerator of F 1 (1) (Dirichlet's integral) does not change in the transition from $\underline{D}$ to $\bar{D}$ and from u to $\bar{u}$. That is, if $d\bar{\sigma}$ is the area-element of $\bar{D}$,

$$\iint_{\underline{D}} |\operatorname{grad} u|^2 \, d\sigma = \iint_{\bar{D}} |\operatorname{grad} \bar{u}|^2 \, d\bar{\sigma} . \tag{1}$$

On the other hand we shall prove that

$$\iint_{\underline{D}} u^2 \, d\sigma \leqq \iint_{\bar{D}} \bar{u}^2 \, d\bar{\sigma} . \tag{2}$$

Also the boundary condition $\bar{u} = 0$ will be satisfied. This will yield the assertion immediately.

We observe that this argument differs from that of Faber and Krahn. They define the "symmetrized" function $\bar{u}$ by the condition $\bar{u} = \rho$ and show that this process diminishes the integral in the numerator and leaves the integral in the denominator unchanged.

(b) In what follows we use the notation

$$G = |\operatorname{grad} u| = d\rho/dn \tag{3}$$

where $d\rho > 0$ and dn is the piece of the normal of the level curve $\underline{C}_\rho$ between the level curves $\underline{C}_\rho$ and $\underline{C}_{\rho + d\rho}$. The contribution of the ring-shaped domain between these curves to Dirichlet's integral is $\int G^2 \, ds.dn = \int G \, ds.d\rho$. Introducing the function

$$P(\rho) = \int_{\underline{C}_\rho} G \, ds \tag{4}$$

we have

$$\iint_{\underline{D}} |\operatorname{grad} u|^2 \, d\sigma = \int_0^1 P(\rho) \, d\rho . \tag{5}$$

The area of the ring-shaped domain mentioned above is $A(\rho) - A(\rho + d\rho) = - A'(\rho) \, d\rho$, or in another form

(6) $$\int d\sigma = \int ds.dn = \int ds.d\rho/G .$$

Consequently

(7) $$-A'(\rho) = |A'(\rho)| = \int_{C_\rho} G^{-1}\, ds.$$

We derive now an important inequality between the quantities introduced before. By Schwarz's inequality

(8) $$\int_{C_\rho} G\, ds \int_{C_\rho} G^{-1}\, ds \geq \left(\int_{C_\rho} ds\right)^2 = [L(\rho)]^2$$

where $L(\rho)$ is the length of C_ρ. Using the isoperimetric inequality (valid also for a system of mutually exclusive curves) $[L(\rho)]^2 \geq 4\pi A(\rho)$ we obtain

(9) $$P(\rho) \geq \frac{4\pi A(\rho)}{|A'(\rho)|}, \quad \rho > 0 .$$

We wish to determine the function $f(\rho)$ so that not only Dirichlet's integral should remain unchanged but even the infinitesimal part of it corresponding to the ring-shaped domain between C_ρ and $C_{\rho + d\rho}$.

(c) Accordingly, we define $\bar{u} = f(\rho)$ as follows:

(10) $$f(\rho) = \int_0^\rho [P(t)]^{1/2} \left[\frac{|A'(t)|}{4\pi A(t)}\right]^{1/2} dt , \quad \rho > 0 .$$

This integral exists since $P(\rho)$ and $A'(\rho)$ are integrable. Also it tends to zero with ρ so that $\bar{u}$ satisfies the boundary condition. We have $A'(\rho)\, d\rho = 2\pi R\, dR$, hence

---

(11) $$|\text{grad}\ \bar{u}|^2 = \left(\frac{d\bar{u}}{dR}\right)^2 = [f'(\rho)]^2 \left(\frac{d\rho}{dR}\right)^2$$
$$= [f'(\rho)]^2 \frac{4\pi^2 R^2}{[A'(\rho)]^2} = [f'(\rho)]^2 \frac{4\pi A(\rho)}{[A'(\rho)]^2} = \frac{P(\rho)}{|A'(\rho)|} .$$

The area between the circles of radii R and $R + dR$ is the same as between C_ρ and $C_{\rho + d\rho}$, that is, $|A'(\rho)|\, d\rho$ so that we find

(12) $$\iint_{\bar{D}} |\text{grad}\ \bar{u}|^2\, d\bar{\sigma} = \int_0^1 P(\rho)\, d\rho$$

which shows that Dirichlet's integral remains unchanged.

On the other hand we conclude from (10) by (9),

(13) $$f(\rho) \geq \int_0^\rho dt = \rho \quad .$$

Hence

(14) $$\int_0^1 \rho^2 \, |A'(\rho)| \, d\rho \leq \int_0^1 [f(\rho)]^2 \, |A'(\rho)| \, d\rho$$

which proves (2). Thus the assertion is established.

We note that (2) remains valid even if $\underline{D}$ is replaced by the infinitesimal ring-shaped domain between the level curves $\underline{C}_\rho$ and $\underline{C}_{\rho + d\rho}$ and $\bar{D}$ by the ring between the circles arising from these curves by symmetrization.

F 5. Clamped plate. (a) We denote the minimizing function of F 1 (2) again by u and assume that $0 \leq u \leq 1$ holds throughout the domain $\underline{D}$. Let $0 < \rho < 1$. We consider the open set $0 < u < \rho$ which consists in general of several simply or multiply connected components. One of them has the curve $\underline{C}$ ($u = 0$) as part of its boundary. We denote the total boundary of $u < \rho$ (without the curve $\underline{C}$) by $\underline{C}_\rho$; it consists of all the curves along which $u = \rho$. Let $\underline{C}'_\rho$ be that part of $\underline{C}_\rho$ which together with $\underline{C}$ forms the boundary of a component of $u < \rho$.

We denote by $A(\rho)$ the area of the complementary set characterized by the condition $u \geq \rho$. The function $A(\rho)$ is continuous and monotonically decreasing, $A(0)$ is the area of the given domain $\underline{D}$ and $A(1) = 0$.

(b) We write again $|\text{grad } u| = G$. There is no need for any essential change in the argument of F 4 (b); in particular the definition F 4 (4) and the inequality F 4 (9) hold. Indeed, let $L(\rho)$ denote the total length of the curves forming $\underline{C}_\rho$. We have then F 4 (8) and we can apply again the isoperimetric inequality $[L(\rho)]^2 \geq 4\pi A(\rho)$. (Obviously more than this is true: $L(\rho)$ can be replaced by the total length of the curves $\underline{C}'_\rho$ and $A(\rho)$ by the total area bounded by the curves of $\underline{C}'_\rho$.) This establishes F 4 (9).

Now we introduce the notation

(1) $$\int_{\underline{C}} (\nabla^2 u)^2 \, G^{-1} \, ds = Q(\rho), \qquad \rho > 0 \; .$$

The meaning of this integral is obvious: $Q(\rho)d\rho$ represents that part of the integral in the numerator of F 1 (2) which corresponds to the infinitesimal ring-shaped domain between $\underline{C}_\rho$ and $\underline{C}_{\rho + d\rho}$. Indeed this is

(2) $$\int (\nabla^2 u)^2 \, d\sigma = \int (\nabla^2 u)^2 \, ds.dn = \int (\nabla^2 u)^2 \, G^{-1} \, ds.d\rho$$

or $Q(\rho) \, d\rho$. By Schwarz's inequality

$$(3)\qquad \int_{C_\rho} (\nabla^2 u)^2 G^{-1}\, ds \int_{C_\rho} G^{-1}\, ds \geqq \left(\int_{C_\rho} (\nabla^2 u) G^{-1}\, ds\right)^2$$

or

$$(4)\qquad [Q(\rho)]^{1/2}\, |A'(\rho)|^{1/2} \geqq \int_{C_\rho} (\nabla^2 u) G^{-1}\, ds .$$

The integral of the left-hand expression exists in $0 \leqq \rho \leqq 1$ since $Q(\rho)$ and $A'(\rho)$ are integrable. Integrating we find

$$\int_0^\rho [Q(t)]^{1/2}\, |A'(t)|^{1/2}\, dt \geqq \int_0^R \int_{C_t} (\nabla^2 u) G^{-1}\, ds.dt$$

$$(5)\qquad = \iint_{0\leqq u\leqq \rho} \nabla^2 u\, d\sigma = - \int_{u=0,\, u=\rho} \frac{\partial u}{\partial n}\, ds = - \int_{C_\rho} \frac{\partial u}{\partial n}\, ds$$

$$= P(\rho) \geqq \frac{4\pi A(\rho)}{|A'(\rho)|} .$$

The integration has to be extended over the set C_ρ and the normal is directed into the interior of the domain $0 < u < \rho$. We used F 4 (3), (4) and finally (9).

(c) We proceed to the definition of $f(\rho)$; this must be given in such a way that the infinitesimal part of the integral under consideration, that is, $Q(\rho)\, d\rho$ should remain unchanged. We set accordingly

$$(6)\qquad g(\rho) = \frac{|A'(\rho)|}{4\pi A(\rho)} \int_0^\rho [Q(t)]^{1/2}\, |A'(t)|^{1/2}\, dt, \quad \rho > 0 ,$$

and

$$(7)\qquad f(\rho) = \int_0^\rho g(\rho)\, d\rho , \quad f(0) = 0 .$$

(The convergence of (7) can be established by integration by parts.) In view of (5) we have

$$(8)\qquad f'(\rho) = g(\rho) \geqq \frac{|A'(\rho)|}{4\pi A(\rho)} P(\rho) \geqq 1$$

so that the inequality

(9) $$f(\rho) \geqq \rho, \quad \rho \geqq 0 ,$$

results.

We determine R as in F 4 so that $A(0) = \pi R_0^2$ where R_0 is the radius of the circle $\bar{D}$ which has the same area as the given domain $\underline{D}$. We consider now the function $\bar{u} = f(\rho)$ defined on $\bar{D}$. It satisfies the boundary conditions

$$(\bar{u})_{R=R_0} = f(0) = 0 ,$$

(10) $$\left(\frac{\partial \bar{u}}{\partial n}\right)_{R=R_0} = \lim_{R \to R_0} \frac{d\bar{u}}{dR}$$

$$= \lim_{\rho \to 0} f'(\rho) \frac{2\pi R}{|A'(\rho)|} = 0$$

in view of (7) and (6). We have moreover

(11) $$\nabla^2 \bar{u} = \frac{1}{R} \frac{d}{dR} \left(R \frac{d\bar{u}}{dR}\right) = \frac{4\pi}{A'(\rho)} \frac{d}{d\rho} \left(\frac{A(\rho) f'(\rho)}{A'(\rho)}\right)$$

$$= \frac{4\pi}{|A'(\rho)|} \frac{d}{d\rho} \left(\frac{A(\rho) f'(\rho)}{|A'(\rho)|}\right) = \frac{[Q(\rho)]^{1/2}}{|A'(\rho)|^{1/2}}$$

and the area of the infinitesimal ring-shaped domain being $|A'(\rho)|\, d\rho$ we find for its contribution to the integral over $\bar{D}$:

(12) $$(\nabla^2 \bar{u})^2 \, |A'(\rho)| \, d\rho = Q(\rho) \, d\rho .$$

Consequently this process does not change the integral in the numerator of F 1 (2).

On the other hand, the integral in the denominator will be for the circular domain $\bar{D}$:

(13) $$\iint_{\bar{D}} \bar{u}^2 \, d\bar{\sigma} = \int_0^1 [f(\rho)]^2 \, |A'(\rho)| \, d\rho$$

$$\geqq \int_0^1 \rho^2 \, |A'(\rho)| \, d\rho .$$

This completes the proof.

The closing remark of F 4 holds again.

F 6. Buckling of a plate. Dealing with problem F 1 (3) the previous argument needs only slight modification. We define $f(\rho)$ in the same way as in F 5 so that the integral in the numerator of F 1 (3) does not change.

As to the integral in the denominator we have by F 4 (5)

$$\iint_{D} |\text{grad } u|^2 \, d\sigma = \int_0^1 P(\rho) \, d\rho \quad . \tag{1}$$

On the other hand (cf. F 5 (10))

$$|\text{grad } \bar{u}|^2 = [f'(\rho)]^2 \frac{4\pi A(\rho)}{[A'(\rho)]^2} \tag{2}$$

and the area between R and R + dR is $|A'(\rho)| \, d\rho$ so that

$$\iint_{\bar{D}} |\text{grad } \bar{u}|^2 \, d\bar{\sigma} = \int_0^1 [f'(\rho)]^2 \frac{4\pi A(\rho)}{|A'(\rho)|} \, d\rho \quad . \tag{3}$$

But by F 5 (8)

$$[f'(\rho)]^2 \frac{4\pi A(\rho)}{|A'(\rho)|} \geqq f'(\rho) \frac{4\pi A(\rho)}{|A'(\rho)|} \geqq P(\rho) \quad . \tag{4}$$

This establishes the desired inequality between (3) and (1).

NOTE G. VIRTUAL MASS AND POLARIZATION*

G 1. Definitions. (a) Let us consider the infinite uniform flow of an incompressible, frictionless, and homogeneous fluid filling the whole space and moving in a direction given by the unit vector $\vec{h}$. By immersing a solid in this fluid, the flow will be disturbed; the disturbance is equivalent to superimposing on the original flow another one. If we denote the velocity potential of the superimposed flow by φ, its energy will be given, apart from trivial factors, by Dirichlet's integral

$$W = \iiint |\operatorname{grad} \varphi|^2 \, d\tau \quad . \tag{1}$$

Here $d\tau$ is the volume element and the integration is extended over the exterior of the solid.

Let us consider an infinite uniform electric field whose direction is determined by the unit vector $\vec{h}$. By placing a conducting solid in this field, the uniform field will be disturbed; the disturbance is equivalent to superimposing on the original field another one. If we denote the electric potential of the superimposed field by Ψ, its energy will be given, apart from trivial factors, by Dirichlet's integral

$$P = \iiint |\operatorname{grad} \Psi|^2 \, d\tau \quad . \tag{2}$$

Again, this integral is extended over the exterior of the given solid.

Both functions φ and Ψ are harmonic and behave like a dipole at infinity, that is, they are of order r^{-2} as $r \to \infty$. But the boundary conditions characterizing them are very different. In the first case we have on the surface A of the given solid B:

$$-\partial \varphi / \partial n = \vec{h} \cdot \vec{n} \tag{3}$$

where $\vec{n}$ is a unit vector in the direction of the exterior normal. In the second case, we have on A:

* Concerning this subject see Pólya (19), Szegö (33), and Schiffer-Szegö (28); also Herriot (11).

$$(4) \qquad \Psi = \vec{h} \cdot \vec{r} + \text{const.}$$

where $\vec{r}$ is the radius vector. (The additive constant must be chosen properly.) The first is a special case of the so-called Neumann problem, the second of the Dirichlet problem.

We recall the analogous case of the capacity C of the solid B; C is defined by the condition that $4\pi C$ is the Dirichlet integral of the harmonic function u which behaves like the potential of a point-charge at infinity and satisfies the condition $u = 1$ on the surface A. It is of order r^{-1} at $r \to \infty$.

As to the energies (1) and (2) they are easily verified to be quadratic forms of the components h_1, h_2, h_3 of $\vec{h}$:

$$(5) \qquad W = \sum_{i,k=1,2,3} W_{ik} h_i h_k , \quad P = \sum_{i,k=1,2,3} P_{ik} h_i h_k .$$

We call W the <u>virtual mass</u> or added mass in the $\vec{h}$-direction and P the <u>polarization</u> in this direction. The coefficients of these forms depend, of course, on the choice of the coordinate system but their invariants are independent of such choice. The simplest invariant is the trace

$$(6) \qquad W_{11} + W_{22} + W_{33} = 3W_m .$$

We call (Pólya 19) the invariant W_m the <u>average virtual mass</u>. In a similar fashion we may introduce the <u>average polarization</u> P_m defined by

$$(7) \qquad P_{11} + P_{22} + P_{33} = 3P_m$$

(Schiffer-Szegö 28). Obviously W_m and P_m are the mean-values of W and P on the unit sphere h_1, h_2, h_3, respectively.

The other invariants of the forms (5) offer also some interest. They are:

$$(8) \qquad \begin{vmatrix} W_{22} & W_{23} \\ W_{32} & W_{33} \end{vmatrix} + \begin{vmatrix} W_{33} & W_{31} \\ W_{13} & W_{11} \end{vmatrix} + \begin{vmatrix} W_{11} & W_{12} \\ W_{21} & W_{22} \end{vmatrix} ; \quad |W_{ik}| \quad i,k = 1,2,3 ,$$

and similarly for the polarization.

(b) The capacity C of the solid B can be interpreted as the point-charge whose potential coincides with the leading term of the harmonic function u which satisfies the condition $u = 1$ on the surface A. In a similar fashion we can relate the virtual mass and polarization to the leading terms of the harmonic functions Φ and Ψ.

Let Φ_1, Φ_2, Φ_3, be the special cases of the harmonic function Φ

corresponding to the special unit vectors along the directions of the coordinate axes x_1, x_2, x_3.

Let the expansion of the harmonic function Φ_i at infinity have the leading term

$$r^{-3}(d_{i1}x_1 + d_{i2}x_2 + d_{i3}x_3) \ . \tag{9}$$

We can show then [Schiffer-Szegö 28, 2.7 (11)] that $d_{ik} = d_{ki}$. We find for the leading term of φ :

$$r^{-3}\sum_{i=1}^{3} h_i(d_{i1}x_1 + d_{i2}x_2 + d_{i3}x_3) \ . \tag{10}$$

A similar expression is obtained for the leading term of Ψ containing the coefficients $e_{ik} = e_{ki}$. The strength-components of the corresponding dipoles in the $\vec{h}$-direction will be given then by the following quadratic forms:

$$D = \sum_{i,k=1,2,3} d_{ik}h_ih_k \ , \quad E = \sum_{i,k=1,2,3} e_{ik}h_ih_k \ . \tag{11}$$

They are called the dipole forms associated with the virtual mass and polarization, respectively.

By the use of Green's formula the following relations can be established (G. I. Taylor 37):

$$W = 4\pi D - V \ , \quad P = 4\pi E - V \tag{12}$$

where V is the volume of the solid B. The quantity D has a simple meaning. it is proportional to the sum of the virtual mass of the solid and the mass of the displaced fluid. The quantities D and E show a more regular behavior than the forms W and P.

The simplest special case is that of a sphere. Denoting the radius by a we have

$$\varphi = \frac{a^3}{2}\frac{\vec{h}\cdot\vec{r}}{r^3} \ , \quad D = \frac{1}{2}a^3(h_1^2 + h_2^2 + h_3^2) = \frac{1}{2}a^3 \ , \quad W = \frac{2\pi}{3}a^3 = \frac{1}{2}V \ , \tag{13}$$

$$\Psi = a^3\frac{\vec{h}\cdot\vec{r}}{r^3} \ , \quad E = a^3(h_1^2 + h_2^2 + h_3^2) = a^3 \ , \quad P = \frac{8\pi}{3}a^3 = 2V \ . \tag{14}$$

Concerning other special cases we refer to Schiffer-Szegö (28). Both the virtual mass and the polarization have the dimension of a volume.

G 2. Problems and results. (a) The principal purpose is to find relations of the constants W_m, P_m, depending only on the given solid B, to geometrical or other data of B. (There is little likelihood of obtaining simple results for the virtual mass or polarization in an arbitrary direction.) In the case of the capacity this aim was achieved by using the principles of Dirichlet and Thomson (ch. 2). They furnish upper and lower bounds for the capacity.

The corresponding "principles" can be formulated as follows:

$$\text{(I)} \qquad W = \max \frac{\left[\iint f\, \vec{h} \cdot \vec{n}\, d\sigma\right]^2}{\iiint |\operatorname{grad} f|^2\, d\tau}$$

where f is an arbitrary function defined in the exterior of B; $d\sigma$ is the element of the boundary surface A of B over which the integral in the numerator is extended; $d\tau$ is the volume element of the exterior of B over which the integral in the denominator is extended.

$$\text{(II)} \qquad W = \min \iiint |\vec{q}|^2\, d\tau$$

where $\vec{q}$ is any vector function defined outside of B, satisfying the conditions:

$$\operatorname{div} \vec{q} = 0 \text{ outside of } B\ ,$$

$$-\vec{q} \cdot \vec{n} = \vec{h} \cdot \vec{n} \text{ on } A\ .$$

For the polarization we have:

$$\text{(III)} \qquad P = \min \iiint |\operatorname{grad} f|^2\, d\tau$$

where f is arbitrary in the exterior of B and satisfies the condition $f = \vec{h} \cdot \vec{r} + \text{const.}$ on A.

(b) An application of the principle (I) is suggested by 2.5. We choose a family of surfaces outside of A defined by the parameter ν, which varies from 0 to ∞. The surface A corresponds to $\nu = 0$ and a large (almost spherical) surface to $\nu = \infty$. We take $f = \vec{h} \cdot \vec{r}\, F(\nu)$ and determine the function F in such a way that the ratio in (I) should be a maximum. This leads to a lower bound for W (and also for W_m) depending only on the chosen family of surfaces.

Specializing this family in various ways we obtain the following lower bounds for the average virtual mass W_m and for the radius $\bar{W}_m$ of the sphere whose average virtual mass is W_m; of course, $W_m = 2\pi(\bar{W}_m)^3/3$. We assume that A is star-shaped with respect to a fixed point in the interior of A; let r be the radius vector drawn from that point and leading to a point of A; let $d\omega$ denote the surface element of the unit sphere at a point where the normal is parallel to the normal of A at the end point of r. We have

(1) $$\bar{W}_m \geqq \left[\frac{1}{4\pi}\iint r^{\lambda}\,d\omega\right]^{1/\lambda}, \quad \lambda \leqq -3,$$

(2) $$W_m \geqq \frac{V}{2}\left[1 + \frac{2}{9V}\iint \nabla(r^{3/2})\,d\omega\right]^{-1}$$

where V is the volume of $\underline{B}$ and $\nabla(u)$ denotes Beltrami's operator on the unit sphere, i.e., the square of the gradient of u on the unit sphere.* Both integrations are extended over the unit sphere.

The quantity on the right of (1), called N_λ in 3.3 (2), is an increasing function of λ. For $\lambda = 3$ it represents the radius $\bar{V}$ of the sphere whose volume is the same as that of $\underline{B}$. Thus inequality (1) is still far from the following one which has been conjectured (Pólya 19):

(3) $$\bar{W}_m \geqq \bar{V} \quad \text{or} \quad W_m \geqq \frac{V}{2}.$$

It would correspond to the inequality of Poincaré (3.2) and could be formulated as follows: Of all solids with given volume the sphere has the minimum average virtual mass.

G 3. Further results. Another interesting application of the extremum principles of G 2 is the following: The dipole forms D and E are monotonically increasing set-functions; that is, D and E are increasing if we replace the given solid $\underline{B}$ by another $\underline{B}'$ which contains $\underline{B}$.

Finally we mention the following inequality connecting the virtual mass and polarization:

(1) $$WP \geqq V^2$$

and

(2) $$W_m P_m \geqq V^2.$$

For the proofs we refer to Schiffer-Szegö (28). As to the average virtual mass and polarization of a nearly spherical solid, cf. Szegö (33) and the table in 1.33. These results speak in favor of the conjecture G 2 (3) and the corresponding conjecture for the polarization.

G 4. Two-dimensional (cylindrical) problems. We consider a cylindrical solid whose cross-section is a closed plane curve $\underline{C}$. The definitions of sect. G 1 have to be modified in this case as follows.

(a) We represent $\underline{C}$ in the complex z-plane, $z = x + iy = re^{i\theta}$, and let $\vec{h}$ be a vector in this plane represented by the complex number $e^{i\alpha}$, α real. We form the velocity potential φ which is harmonic in the exterior of $\underline{C}$, has the form

* See W. Blaschke, Vorlesungen über Differentialgeometrie, p. 113.

(1) $$\varphi = \frac{d_1 \cos\theta + d_2 \sin\theta}{r} + \frac{d_1' \cos 2\theta + d_2' \sin 2\theta}{r^2} + \dots$$

at infinity and satisfies the boundary condition

(2) $$-\partial\varphi/\partial n = \vec{h} \cdot \vec{n} \text{ on } \underline{C}.$$

The virtual mass of $\underline{C}$ in the direction $\vec{h}$ is given by

(3) $$W = \iint |\text{grad}\ \varphi|^2\, dx\, dy ,$$

the integration being extended over the exterior of $\underline{C}$.

The polarization is defined in an analogous manner by considering a harmonic function of the form

(4) $$\Psi = \frac{e_1 \cos\theta + e_2 \sin\theta}{r} + \frac{e_1' \cos 2\theta + e_2' \sin 2\theta}{r^2} + \dots$$

at infinity and satisfying the condition

(5) $$\Psi = \vec{h} \cdot \vec{r} + \text{const. on } \underline{C} .$$

Then

(6) $$P = \iint |\text{grad}\ \Psi|^2\, dx\, dy ,$$

integrating as before.

We introduce also the strength components in the $\vec{h}$-direction

(7) $$D = d_1 \cos\alpha + d_2 \sin\alpha , \quad E = e_1 \cos\alpha + e_2 \sin\alpha .$$

(b) The quantities defined above can be computed by using the mapping function

(8) $$z = \bar{r}(\zeta + c_0 + c_1\zeta^{-1} + c_2\zeta^{-2} + \dots)$$

where $|\zeta| > 1$ and the exterior of $\underline{C}$ correspond to each other; $\zeta = \infty$ is transformed into $z = \infty$. The coefficient $\bar{r}$ is the outer radius of $\underline{C}$. We then have

(9) $$\varphi = \Re\left\{e^{-i\alpha}\bar{r}(\zeta + e^{2i\alpha}\zeta^{-1} + c_0) - e^{-i\alpha} z\right\} ,$$

(10) $$\Psi = \Re\left\{-e^{-i\alpha}\bar{r}(\zeta - e^{2i\alpha}\zeta^{-1} + c_0) + e^{-i\alpha} z\right\} .$$

For the virtual mass and polarization the following formulas hold:

(11) $$W = -A + 2\pi\bar{r}^2\,[1 - \Re(c_1 e^{-2i\alpha})] ,$$

(12) $$P = -A + 2\pi\bar{r}^2 [1 + \Re (c_1 e^{-2i\alpha})] ,$$

where

(13) $$A = \pi\bar{r}^2 (1 - |c_1|^2 - 2|c_2|^2 - 3|c_3|^2 - \ldots)$$

is the area of the closed interior of the curve $\underline{C}$.

(c) For the proof of (9) we note that the function on the right of (9) is of order r^{-1} as $r \to \infty$, and that the condition (2) can be written in the following form:

(14) $$\frac{\partial}{\partial n} (\varphi + \vec{h} \cdot \vec{r}) = 0 ;$$

here $\vec{r}$ is the vector leading from the origin to the point $z = x + iy$ so that $\vec{h} \cdot \vec{r} = \Re(e^{-i\alpha} z)$. Let $\zeta = \rho e^{i\omega}$. Condition (14) is equivalent to

$$\frac{\partial}{\partial \rho} [\varphi + \Re (e^{-i\alpha} z)] = 0 ,$$

or

$$\Re \frac{\partial}{\partial \rho} [e^{-i\alpha}\bar{r}(\zeta + e^{2i\alpha} \zeta^{-1} + c_0)]$$

$$= \Re \bar{r}(\zeta e^{-i\alpha} - \zeta^{-1} e^{i\alpha}) = 0 , \quad |\zeta| = \rho = 1 .$$

The last equation is trivial. An even simpler calculation yields (10). As to (11) and (12) we note that Dirichlet's integral is invariant under conformal mapping; hence

$$W = \int_1^{\infty}\int_{-\pi}^{\pi} |e^{-i\alpha}\bar{r}(1 - e^{2i\alpha} \zeta^{-2}) - e^{-i\alpha} \frac{dz}{d\zeta}|^2 \rho \, d\omega \, d\rho$$

$$= \bar{r}^2 \int_1^{\infty}\int_{-\pi}^{\pi} |(c_1 - e^{2i\alpha}) \zeta^{-2} + 2c_2 \zeta^{-3} + 3c_3 \zeta^{-4} + \ldots|^2 \rho \, d\omega \, d\rho$$

$$= 2\pi\bar{r}^2 \int_1^{\infty} \left\{|c_1 - e^{2i\alpha}|^2 \rho^{-4} + 4|c_2|^2 \rho^{-6} + 9|c_3|^2 \rho^{-8} + \ldots\right\} \rho \, d\rho$$

$$= \pi\bar{r}^2 (|c_1 - e^{2i\alpha}|^2 + 2 |c_2|^2 + 3 |c_3|^2 + \ldots)$$

which is equivalent to (11). We obtain in a similar fashion (12). The strength components are:

(15) $$D = \bar{r}^2 [1 - \Re (c_1 e^{-2i\alpha})] , \quad E = \bar{r}^2 [1 + \Re (c_1 e^{-2i\alpha})]$$

so that, corresponding to G 1 (12),

(16) $$W = 2\pi D - A\ ,\quad P = 2\pi E - A\ .$$

We obtain from (11) and (12) for the average virtual mass and average polarization (mean values of W and P for all α, $-\pi < \alpha \leqq \pi$)

(17) $$W_m = P_m = -A + 2\pi \bar{r}^2\ .$$

In the special case of a circle of radius a we have $W = P = \pi a^2 = A$. We obtain from (17) in view of the area theorem $A \leqq \pi \bar{r}^2$, that

(18) $$W_m \geqq A\ ,\quad P_m \geqq A\ ;$$

i.e., <u>of all domains of a given area the circle has the least average virtual mass and the least average polarization</u>.

This is another instance in favor of the conjecture G 2 (3). Cf. Pólya (<u>19</u>), Szegö (<u>33</u>, pp. 222-223), and Schiffer-Szegö (<u>28</u>, pp. 203-205).

TABLES

FOR SOME SET-FUNCTIONS OF PLANE DOMAINS

1. These tables deal with the following ten quantities (set-functions):

L length of perimeter,
A area,
I moment of inertia of the area with respect to its centroid,
B minimum of $\int h^{-1} ds$, see 1.20 (b),
ρ radius of the inscribed circle, see 5.11 (c),
R radius of the circumscribed circle, see 5.11 (c),
$\dot{r}$ maximum inner radius,
$\bar{r}$ outer radius,
P torsional rigidity,
Λ principal frequency.

We refer to 1.3 for those seven quantities for which no reference is given above. The first three quantities are defined by definite integrals, the fourth by a minimum problem concerned with a definite integral, the next two by minimum problems, and the last four by boundary value problems for partial differential equations. The first six may be called "geometrical quantities", the last four "physical quantities".

The first thirteen tables list expressions or numerical values of these ten quantities for the following thirteen domains:

Circle	Ellipse	Narrow ellipse
Square	Rectangle	Narrow rectangle
Semicircle	Sector	Narrow sector
Triangle $60^{\circ}, 60^{\circ}, 60^{\circ}$	Triangle $45^{\circ}, 45^{\circ}, 90^{\circ}$	Triangle $30^{\circ}, 60^{\circ}, 90^{\circ}$
	Regular hexagon	

In certain tables a few quantities are missing since their values are not sufficiently known. The symbols for the characteristic data of each domain are explained in the lines following the title of the corresponding table. These symbols are retained in the sequel.

The following fourteen tables deal with fourteen different products of powers of the ten quantities:

$$\dot{r}A^{-1/2}, \qquad \bar{r}A^{-1/2}, \qquad \bar{r}L^{-1},$$

$$PA^{-2}, \qquad PA^{-2}B, \qquad PA^{-4}I, \qquad P\dot{r}^{-4},$$

$$\Lambda A^{1/2}, \qquad \Lambda A^{1/2}B^{-1/2}, \qquad \Lambda A^{3/2}I^{-1/2}, \qquad \Lambda\dot{r},$$

$$P\Lambda^{4}, \qquad P\Lambda^{2}A^{-1}, \qquad P\bar{r}^{4}A^{-4}.$$

Each of these products of powers depends only on the shape of the domain and is independent of its size (is of dimension 0); each is connected with an inequality or an approximate formula. A short remark at the end of the table summarizes some important feature suggested or confirmed by the table. The term "bounded" must be taken here in the following sense: the product of powers in question remains between two finite positive bounds for all convex plane domains. The ratio of the two extreme values listed in the table is given whenever it is finite (since it can help to judge the usefulness of the corresponding approximation or inequality). A reference to one or more pertinent sections is added.

Two more tables list expressions for B, $\dot{r}$, and $\bar{r}$ for a few additional domains, different from the above thirteen.

The annotations to all tables are consecutively numbered and collected at the end. They explain the notation for certain functions and constants and quote sources for more recent material not treated in the usual textbooks.

2. The study of these tables suggests interesting observations. We confine ourselves to two remarks.

(a) Some of the tabulated products of powers behave similarly as L^2A^{-1}, the "isoperimetric quotient". The main property of L^2A^{-1} is that it attains its minimum when the domain is a circle. Yet it has several other properties of which we quote a few.

I. Of all triangles, the equilateral triangle attains the minimum value of L^2A^{-1}.

II. Of all quadrilaterals, the square attains the minimum value of L^2A^{-1}.

III. As the number of the sides of a regular polygon increases, L^2A^{-1} decreases.

IV. As the ratio of the major axis of an ellipse to its minor axis decreases, also L^2A^{-1} decreases (and attains its minimum for the circle).

V. As the ratio of the length of a rectangle to its width decreases, also L^2A^{-1} decreases (and attains its minimum for the square).

VI. Let γ be the opening of a sector of the circle. There is a value γ_0, $0 < \gamma_0 < 2\pi$, such that L^2A^{-1} decreases when γ increases from 0 to γ_0 and L^2A^{-1} increases when γ increases from γ_0 to 2π (and so L^2A^{-1} attains its minimum for $\gamma = \gamma_0$).

VII. If an angle of a triangle is equal to the opening of a sector of the circle, L^2A^{-1} is less for the sector than for the triangle.

Each of the following nine products of powers

$$\dot{r}^{-1}A^{1/2}, \qquad \bar{r}A^{-1/2}, \qquad \bar{r}^{-1}L,$$

$$P^{-1}A^{2}, \qquad PA^{-2}B, \qquad P\dot{r}^{-4},$$

$$\Lambda A^{1/2}, \qquad \Lambda^{-1}A^{-1/2}B^{1/2}, \qquad \Lambda^{-1}\dot{r}^{-1}$$

is analogous to the isoperimetric quotient L^2A^{-1} in so far as each attains its minimum when the domain is a circle. There are some reasons to suspect that also

$$P\Lambda^4$$

behaves in the same manner. Analogy may suggest that the ten products mentioned share with L^2A^{-1} also the properties I-VII. This means, in fact, seventy different conjectures. Many of these have been proved (see especially 7.4), none of them is contradicted by the tables (if we take the term "decreasing" in the wide sense), and it seems not unreasonable to expect that an overwhelming majority will turn out to be actually true.

(b) The domains considered in the thirteen initial tables are all convex with one exception: a sector of a circle is not convex if its opening exceeds π. The consideration of the quantity B and of any product into which B enters may be restricted to convex domains and must be restricted to star-shaped domains. Yet in the other cases it may be misleading to confine the consideration to convex domains. In order to obtain an unbiased orientation we have to construct rather artificial non-convex domains of which we indicate two examples.

We take away from a circular disk narrow angular neighborhoods of n equidistant radii, but restitute those parts which belong to the interior of a smaller concentric circle. We obtain so a domain F_n (the "flower") star-shaped with respect to the center and bounded by a closed curve that consists of alternating radial segments and concentric circular arcs. We choose the opening of the angular neighborhoods and the radius of the concentric circle as very rapidly decreasing functions of n. Then we can treat F_n for large n, at least in certain respects, as an assemblage of n equal (virtually separated) sectors.

As a second example we consider the domain D_n (the "dumbbell") obtained as the union of three domains: two circles and a rectangle. The length of the long side of the rectangle is n, the length of the short side is a very rapidly decreasing function of n. The circles have the same radius 1, and the center of each is the midpoint of a short side of the rectangle. For large n we can treat D_n in some respects as the assemblage of two widely separated disks.

For both domains we can compute with sufficient approximation the quantities

$$L, \quad A, \quad I, \quad \rho, \quad R, \quad \dot{r}, \quad \bar{r}, \quad P, \quad \Lambda$$

which turn out to be asymptotically proportional to the following powers of n:

$$n, \quad 1, \quad 1 \quad n^{-1}, \quad 1, \quad n^{-1}, \quad 1, \quad n^{-2}, \quad n$$

$$n, \quad 1, \quad n^{2}, \quad 1 \quad n, \quad 1, \quad n, \quad 1, \quad 1$$

in the case of F_n and in that of D_n, respectively.

What the entries 12, 19 and 20 of Table 1.21 assert about lower bounds 0 and upper bounds ∞ for non-convex figures can be verified by use of the results just listed, which we can also use in several other instances in constructing counter-examples.

CIRCLE

Radius a

$$L = 2\pi a$$
$$A = \pi a^2$$
$$I = \frac{\pi}{2} a^4$$
$$B = 2\pi$$
$$\rho = a$$
$$R = a$$
$$\dot{r} = a$$
$$\bar{r} = a$$
$$P = \frac{\pi}{2} a^4$$
$$\Lambda = ja^{-1} = 2.4048\ a^{-1} \qquad 1).$$

ELLIPSE

Semi-axes a and b, numerical eccentricity ε ;
$a \geq b \geq 0$, $\varepsilon^2 = (a^2 - b^2)/a^2$.

$$L = 4aE(\varepsilon) \qquad 2)$$
$$A = \pi ab$$
$$I = \frac{\pi}{4} ab(a^2 + b^2)$$
$$B = \pi\left(\frac{a}{b} + \frac{b}{a}\right)$$
$$\rho = b$$
$$R = a$$
$$\log \dot{r} = \log \bar{r} - \sum_{n=1}^{\infty} \frac{2}{n} \frac{q^n}{1 + q^n} \text{ with } q = \left(\frac{a - b}{a + b}\right)^2 \qquad 3)$$
$$\dot{r} = \bar{r}\left\{\sum_{n=0}^{\infty} q^{n(n+1)}\right\}^{-1} \left\{1 + 2\sum_{n=1}^{\infty} q^{n^2}\right\}^{-1}$$
$$\bar{r} = \frac{a + b}{2}$$
$$P = \frac{\pi a^3 b^3}{a^2 + b^2}$$

NARROW ELLIPSE

$$\frac{b}{a} \longrightarrow 0, \qquad \varepsilon \longrightarrow 1$$

$$L \sim 4a$$

$$A = \pi ab$$

$$I \sim \frac{\pi}{4} a^3 b$$

$$B \sim \pi ab^{-1}$$

$$\rho = b$$

$$R = a$$

$$\dot{r} \sim \frac{4}{\pi} b$$

$$\bar{r} \sim \frac{1}{2} a$$

$$P \sim \pi ab^3$$

$$\Lambda \sim \frac{\pi}{2} b^{-1}$$

SQUARE

Side a

$$L = 4a$$

$$A = a^2$$

$$I = 6^{-1} a^4$$

$$B = 8$$

$$\rho = 2^{-1} a$$

$$R = 2^{-1/2} a \qquad = 0.7071\ a$$

$$\dot{r} = \frac{4\pi^{1/2}}{[\Gamma(1/4)]^2}\, a = 0.53935\ a$$

$$\bar{r} = \frac{[\Gamma(1/4)]^2}{4\pi^{3/2}}\, a = 0.59017\ a$$

$$P = 0.1406\ a^4$$

$$\Lambda = 2^{1/2} \pi a^{-1}$$

RECTANGLE

Sides a, b; $a \geq b$.

$$L = 2(a + b)$$

$$A = ab$$

$$I = \frac{1}{12} ab(a^2 + b^2)$$

$$B = 4\left(\frac{a}{b} + \frac{b}{a}\right)$$

$$\rho = \frac{1}{2} b$$

$$R = \frac{1}{2}(a^2 + b^2)^{1/2}$$

$$\dot{r} = \frac{2}{\pi} b \left(1 + 2 \sum_{n=1}^{\infty} q^{n^2}\right)^{-2}, \quad q = e^{-\pi a/b}$$

$$\Lambda = \pi \frac{(a^2 + b^2)^{1/2}}{ab}$$

a/b	$Pa^{-1}b^{-3}$
1	.1406
2	.2287
3	.2633
4	.2808
5	.2913
6	.2983
7	.3033
8	.3071
10	.3123
12	.3158
100	.3312
∞	.3333 = 1/3

NARROW RECTANGLE

$b/a \to 0$

$L \sim 2a$

$A = ab$

$I \sim \frac{1}{12} a^3 b$

$B \sim 4ab^{-1}$

$\rho = \frac{1}{2} b$

$R \sim \frac{1}{2} a$

$\dot{r} \sim \frac{2}{\pi} b$

$\bar{r} \sim \frac{1}{4} a$

$P \sim \frac{1}{3} ab^3$

$\Lambda \sim \pi b^{-1}$

SEMICIRCLE

One half of circle with radius a.

$L = 5.14159\ a$
$A = 1.57080\ a^2$
$I = 0.50246\ a^4$
$B = 8.7915$
$\rho = 0.50000\ a$
$R = 1.00000\ a$
$\dot{r} = 0.60057\ a$
$\bar{r} = 0.76980\ a$
$P = 0.29756\ a^4$
$\Lambda = 3.83171\ a^{-1}$

SECTOR

Radius a, angle (opening) $\gamma = 2\pi\lambda$.
For $\lambda = 1/2,\ 1/4,\ 1/6$ semicircle, quadrant, sextant, respectively.

$$L = (2 + 2\pi\lambda)a$$

$$A = \pi\lambda a^2$$

$$I = \left(\frac{\pi\lambda}{2} - \frac{4}{9}\,\frac{\sin^2\pi\lambda}{\pi\lambda}\right)a^4$$

$$\rho = \begin{cases} \dfrac{\sin\pi\lambda}{1+\sin\pi\lambda}\,a\,, & 0 \leqq \lambda \leqq 1/2 \\ \dfrac{1}{2}a\,, & 1/2 \leqq \lambda \leqq 1 \end{cases}$$

$$R = \begin{cases} \dfrac{a}{2\cos\pi\lambda}\,, & 0 \leqq \lambda \leqq 1/4 \\ a\sin\pi\lambda\,, & 1/4 \leqq \lambda \leqq 1/2 \\ a\,, & 1/2 \leqq \lambda \leqq 1 \end{cases}$$

$$\dot{r} = 4\lambda[(1+\lambda^2)^{1/2} - \lambda]\,[(1+\lambda^{-2})^{1/2} - \lambda^{-1}]^{\lambda}a$$

$$P = \left\{\frac{\pi\lambda}{2} - \frac{1}{\pi}\,[\psi(\tfrac{1}{2} + 2\lambda) - \psi(\tfrac{1}{2}) - \lambda\,\dot{\psi}'(\tfrac{1}{2} + 2\lambda)]\right\} a^4 \qquad 4)$$

$$\Lambda = j_{1/(2\lambda)}a^{-1} \qquad 5)$$

λ	$\dot{r}a^{-1}$	Pa^{-4}	$\Lambda a = j_{1/(2\lambda)}$
1/12	.2353	.006773	9.9361095
1/10	.2682	.01063	8.7714838
1/8	.3120	.01812	7.5883427
1/6	.3728	.03490	6.3801619
1/4	.4625	.08233	5.1356223
1/3	.5242	.14421	4.4934095
1/2	.6006	.29756	3.8317060
3/4	.6580	.57248	
5/6	.6695	.67174	
1	.6863	.87806	3.1415927

NARROW SECTOR

$$a = 1, \quad \gamma = 2\pi\lambda \rightarrow 0$$

$$L = 2 + \gamma$$

$$A = \frac{\gamma}{2}$$

$$I = \frac{1}{36}\gamma + \frac{1}{54}\gamma^3 + O(\gamma^5)$$

$$B = \frac{4}{\gamma} + 4 + O(\gamma)$$

$$\rho = \frac{\gamma}{2}\left[1 - \frac{\gamma}{2} + \frac{5}{24}\gamma^2 - \frac{1}{12}\gamma^3 + O(\gamma^4)\right]$$

$$R = \frac{1}{2} + \frac{\gamma^2}{16} + O(\gamma^4)$$

$$\dot{r} = 4\lambda^{1+\lambda}[1 - (1 + \log 2)\lambda + O(\lambda^2)]$$

$$P = \gamma^3\left[\frac{1}{12} - \frac{31}{\pi^5}\zeta(5)\gamma + \frac{1}{10}\gamma^2 + O(\gamma^3)\right] \qquad 6)$$

$$\Lambda \sim \frac{\pi}{\gamma}$$

EQUILATERAL TRIANGLE

Side a, altitude $h = 3^{1/2}2^{-1}a$.

$$L = 3a$$

$$A = \frac{ah}{2}$$

$$I = \frac{a^3h}{24}$$

$$B = 3^{1/2}6 = 10.392$$

$$\rho = \frac{h}{3}$$

$$R = \frac{2h}{3}$$

$$r = \frac{2\pi a}{[\Gamma(1/3)]^3} = 0.3774\ h$$

$$\bar{r} = \frac{[\Gamma(1/3)]^3 h}{4\pi^2} = 0.4870\ h$$

$$P = \frac{ah^3}{30}$$

$$\Lambda = \frac{2\pi}{h}$$

TRIANGLE 45°, 45°, 90°

Isosceles right triangle with hypotenuse $2^{1/2}a$.

$$L = (2 + 2^{1/2})a$$

$$A = \frac{a^2}{2}$$

$$I = \frac{a^4}{18}$$

$$B = 6 + 2^{1/2}4 = 11.657$$

$$\rho = (1 - 2^{-1/2})a = 0.2929\ a$$

$$R = 2^{-1/2}a = 0.7071\ a$$

$$\dot{r} = \frac{4(2\pi)^{1/2}}{3^{3/4}[\Gamma(1/4)]^2}\,a = 0.33462\ a$$

$$\bar{r} = \frac{3^{3/4}[\Gamma(1/4)]^2}{4(2\pi)^{3/2}}\,a = 0.47563\ a$$

$$P = 0.026091\ a^4$$

$$\Lambda = \pi\, 5^{1/2}a^{-1} = 7.0248\ a^{-1}$$

TRIANGLE 30°, 60°, 90°

Sides a, a/2 and $h = 3^{1/2}2^{-1}a$.

$$L = (1 + 3^{1/2})h$$

$$A = \frac{ah}{4}$$

$$I = \frac{a^3h}{72}$$

$$B = 2(3 + 3^{1/2}2) = 12.928$$

$$\rho = 6^{-1}(3 - 3^{1/2})h = 0.2113\ h$$

$$R = 2^{-1}a \qquad = 0.5774\ h$$

$$\dot{r} = \frac{2^{4/3}\pi a}{5^{5/12}[\Gamma(1/3)]^3} = 0.24315\ h$$

$$\bar{r} = \frac{5^{5/12}[\Gamma(1/3)]^3 h}{2^{10/3}\pi^2} = 0.37791\ h$$

$$P = 0.0044517\ a^4 = 0.0079141\ h^4 \qquad 7)$$

$$\Lambda = 3^{-1}7^{1/2}4\pi a^{-1} = 11.082\ a^{-1} \qquad 8)$$

REGULAR HEXAGON

Side a.

$$L = 6a$$

$$A = 3^{3/2}2^{-1}a^2 = 2.598\ a^2$$

$$I = 2^{-3}5.3^{1/2}a^4 = 1.0825\ a^4$$

$$B = 3^{1/2}4 = 6.9282$$

$$\rho = 2^{-1}3^{1/2}a = 0.8660\ a$$

$$R = a$$

$$\dot{r} = \frac{2^{5/3}3^{1/2}\pi}{[\Gamma(1/3)]^3}a = 0.89850\ a$$

$$\bar{r} = \frac{3[\Gamma(1/3)]^3}{2^{8/3}\pi^2}a = 0.92039\ a$$

$$P = 1.0359\ a^4 \qquad 9)$$

Domain	$\dot{r}A^{-1/2}$
Circle	$0.56419 = \pi^{-1/2}$
Regular hexagon	0.55744
Square	0.53935
Equilateral triangle	0.49665
Ellipse a/b = 2	0.48351
Semicircle	0.47919
Triangle 45°, 45°, 90°	0.47320
Triangle 30°, 60°, 90°	0.45251
Narrow ellipse	$\sim \frac{4}{\pi^{3/2}} (\frac{b}{a})^{1/2} = .7183 (\frac{b}{a})^{1/2}$
Narrow rectangle	$\sim \frac{2}{\pi} (\frac{b}{a})^{1/2} = .6366 (\frac{b}{a})^{1/2}$
Narrow sector	$\sim \frac{2^{3/2}}{\pi} \gamma^{1/2} = .9003 \gamma^{1/2}$

Sector

λ	$\dot{r}A^{-1/2}$
1/12	.4599
1/10	.4785
1/8	.4979
1/6	.5152
1/4	.5219
1/3	.5123
1/2	.4792
3/4	.4287
5/6	.4138
1	.3872

Maximum for circle. Cf. A 6 (c).

$\bar{r}A^{-1/2}$

Domain	$\bar{r}A^{-1/2}$
Circle	$0.56419 = \pi^{-1/2}$
Regular hexagon	0.57101
Square	0.59017
Ellipse a/b = 2	0.59841
Semicircle	0.61421
Equilateral triangle	0.64091
Triangle 45°, 45°, 90°	0.67268
Triangle 30°, 60°, 90°	0.70343
Ellipse	$\frac{1}{2\pi^{1/2}}[(\frac{a}{b})^{1/2} + (\frac{b}{a})^{1/2}]$
Narrow rectangle	$\sim \frac{1}{4}(\frac{a}{b})^{1/2}$

Minimum for circle. Cf. A 6 (c).

$\bar{r}L^{-1}$

Domain	$\bar{r}L^{-1}$
Circle	$0.15915 = 1/(2\pi)$
Ellipse a/b = 2	0.15482
Regular hexagon	0.15340
Semicircle	0.14972
Square	0.14754
Equilateral triangle	0.14058
Triangle 45°, 45°, 90°	0.13932
Triangle 30°, 60°, 90°	0.13833
Narrow ellipse	$\sim$ 0.12500
Narrow rectangle	$\sim 0.12500 = 1/8$

Maximum for circle. Bounded; ratio of extreme values listed: $1.273 = 4/\pi$. Cf. 7.6 (b).

Domain	$2\pi PA^{-2}$
Circle	1.0000
Regular hexagon	0.9643
Square	0.8834
Ellipse a/b = 2	0.8000
Semicircle	0.7577
Equilateral triangle	0.7255
Triangle 45°, 45°, 90°	0.6557
Triangle 30°, 60°, 90°	0.5967
Ellipse	$\frac{2ab}{a^2 + b^2}$
Narrow rectangle	$\sim \frac{2\pi}{3}\frac{b}{a} = 2.0944\,\frac{b}{a}$
Narrow sector	$\sim \frac{2\pi}{3}\gamma = 2.0944\,\gamma$

Rectangle

a/b	$2\pi PA^{-2}$
2	.718
3	.551
4	.441
5	.366
6	.312
7	.272
8	.241
10	.196
12	.165
100	.021

Sector

λ	$2\pi PA^{-2}$
1/12	.6209
1/10	.6767
1/8	.7383
1/6	.7998
1/4	.8386
1/3	.8263
1/2	.7577
3/4	.6479
5/6	.6158
1	.5590

Maximum for circle. Cf. 5.17, A 6 (b).

Domain	$PA^{-2}B$
Ellipse (any shape)	1.000
Semicircle	1.060
Regular hexagon	1.063
Square	1.125
Equilateral triangle	1.200
Triangle 45°, 45°, 90°	1.217
Triangle 30°, 60°, 90°	1.228
Narrow rectangle	~1.333
Narrow sector	~1.333 = 4/3

Rectangle

a/b	$PA^{-2}B$
2	1.145
3	1.168
4	1.195
5	1.213
6	1.227
7	1.235
8	1.248
10	1.263
12	1.274
100	1.325

Minimum for ellipse. Bounded; ratio of extreme values listed: 1.333 = 4/3. Cf. 5.5.

$PA^{-4}I$

Domain	$4\pi^2 PA^{-4}I$
Equilateral triangle	$0.877 = 4\pi^2/45$
Triangle 45°, 45°, 90°	0.916
Square	0.925
Triangle 30°, 60°, 90°	0.962
Semicircle	0.970
Regular hexagon	0.972
Ellipse (any shape)	1.000
Narrow rectangle	~ 1.097
Narrow sector	$\sim 1.462 = 4\pi^2/27$

Rectangle		Sector	
a/b	$4\pi^2 PA^{-4}I$	λ	$4\pi^2 PA^{-4}I$
2	.941	1/12	.978
3	.962	1/10	.947
4	.982	1/8	.920
5	.997	1/6	.909
6	1.009	1/4	.937
7	1.018	1/3	.972
8	1.026	1/2	.970
10	1.038	3/4	.795
12	1.046	5/6	.715
100	1.090	1	.559

Bounded; ratio of extreme values listed: 1.667 = 5/3. Cf. 1.17, 5.11 (a).

$\dot{P}r^{-4}$

Domain	$2\pi^{-1}\dot{P}r^{-4}$
Circle	1.000
Regular hexagon	1.011
Square	1.058
Equilateral triangle	1.209
Triangle 45°, 45°, 90°	1.325
Triangle 30°, 60°, 90°	1.441
Semicircle	1.456
Ellipse $a/b = 2$	1.483
Narrow ellipse	$\sim \frac{\pi^4}{128}\frac{a}{b} = 0.7610\,\frac{a}{b}$
Narrow rectangle	$\sim \frac{\pi^3}{24}\frac{a}{b} = 1.2919\,\frac{a}{b}$
Narrow sector	$\sim \frac{\pi^3}{96}\frac{1}{\gamma} = 0.3230\,\frac{1}{\gamma}$

Sector

λ	$2\pi^{-1}\dot{P}r^{-4}$
1/12	1.407
1/10	1.310
1/8	1.217
1/6	1.150
1/4	1.146
1/3	1.216
1/2	1.456
3/4	1.944
5/6	2.129
1	2.520

Minimum for circle. Cf. 5.7.

$\Lambda A^{1/2}$

Domain	$\pi^{-1} j^{-2} \Lambda^2 A$
Circle	1.0000
Square	1.0865
Equilateral triangle	1.2546
Semicircle	1.2694
Triangle 45°, 45°, 90°	1.3581
Triangle 30°, 60°, 90°	1.4637
Rectangle	$\frac{\pi}{j^2}\left(\frac{a}{b} + \frac{b}{a}\right) = .5432 \left(\frac{a}{b} + \frac{b}{a}\right)$
Narrow ellipse	$\sim \frac{\pi^2}{4j^2} \frac{a}{b} = .4267 \frac{a}{b}$
Narrow sector	$\sim \frac{\pi}{2j^2} \frac{1}{\gamma} = .2716 \frac{1}{\gamma}$

Sector

λ	$\pi^{-1} j^{-2} \Lambda^2 A$
1/12	1.4226
1/10	1.3304
1/8	1.2446
1/6	1.1732
1/4	1.1401
1/3	1.1638
1/2	1.2694
1	1.7066

Minimum for circle. Cf. A 6 (a).

$\Lambda A^{1/2}B^{-1/2}$

Domain	$2j^{-2}\Lambda^2 AB^{-1}$
Circle	1.0000
Semicircle	0.9072
Rectangle (any shape)	$0.8533 = \pi^2 j^{-2}/2$
Narrow ellipse	~ 0.8533
Equilateral triangle	0.7585
Triangle 45°, 45°, 90°	0.7320
Triangle 30°, 60°, 90°	0.7114
Narrow sector	$\sim 0.4267 = \pi^2 j^{-2}/4$

Maximum for circle. Bounded; ratio of extreme values listed: $(1.531)^2 = 4j^2\pi^{-2}$. Cf. 5.6.

$\Lambda A^{3/2} I^{-1/2}$

Domain	$(12)^{-1} \pi^{-2} \Lambda^2 A^3 I^{-1}$
Rectangle (any shape)	1.0000
Equilateral triangle	1.0000
Circle	0.9638
Semicircle	0.9562
Triangle 45°, 45°, 90°	0.9375
Triangle 30°, 60°, 90°	0.8750
Narrow ellipse	$\sim$ 0.8225
Narrow sector	$\sim$ 0.3750 = 3/8

Sector

λ	$(12)^{-1} \pi^{-2} \Lambda^2 A^3 I^{-1}$
1/12	.8706
1/10	.9161
1/8	.9620
1/6	.9949
1/4	.9830
1/3	.9536
1/2	.9562
1	1.6449

Bounded; ratio of extreme values listed: $(1.633)^2 = 8/3$. Cf. 1.19.

$\Lambda \dot{r}$

Domain	$j^{-1}\Lambda \dot{r}$
Circle	1.0000
Square	0.9964
Equilateral triangle	0.9860
Triangle 45°, 45°, 90°	0.9774
Triangle 30°, 60°, 90°	0.9704
Semicircle	0.9570
Narrow ellipse	$\sim$ 0.8317
Narrow rectangle	$\sim$ 0.8317
Narrow sector	$\sim$ 0.8317 $= 2j^{-1}$

Sector

λ	$j^{-1}\Lambda \dot{r}$
1/12	.9722
1/10	.9783
1/8	.9845
1/6	.9891
1/4	.9877
1/3	.9795
1/2	.9570
1	.8966

Maximum for circle. Bounded; ratio of extreme values listed: 1.2024 = $j/2$. Cf. 5.8.

$P\Lambda^4$

Domain	$P\Lambda^4$
Circle	$52.53 = \pi j^4/2$
Square	54.78
Equilateral triangle	59.99
Triangle 45°, 45°, 90°	63.54
Semicircle	64.14
Triangle 30°, 60°, 90°	67.15
Narrow ellipse	$\sim \frac{\pi^5}{16}\frac{a}{b} = 19.13\,\frac{a}{b}$
Narrow rectangle	$\sim \frac{\pi^4}{3}\frac{a}{b} = 32.47\,\frac{a}{b}$
Narrow sector	$\sim \frac{\pi^4}{12}\frac{1}{\gamma} = 8.117\,\frac{1}{\gamma}$

Rectangle

a/b	$P\Lambda^4$
2	69.62
3	94.99
4	123.5
5	153.5
6	184.2
7	215.3
8	246.9
10	310.3
12	374.3
100	3227.

Sector

λ	$P\Lambda^4$
1/12	66.01
1/10	62.93
1/8	60.08
1/6	57.83
1/4	57.27
1/3	58.79
1/2	64.14
1	85.52

Minimum for circle <u>conjectured</u>. Cf. 1.35 (d).

Domain	$P\Lambda^2A^{-1}$
Narrow sector	$\sim 1.6449 = \pi^2/6$
Narrow ellipse	~ 2.4674
Triangle 30°, 60°, 90°	2.5254
Triangle 45°, 45°, 90°	2.5751
Equilateral triangle	2.6319
Square	2.7753
Semicircle	2.7812
Circle	2.8916
Narrow rectangle	$\sim 3.2899 = \pi^2/3$

Rectangle

a/b	$P\Lambda^2A^{-1}$
2	2.822
3	2.887
4	2.945
5	2.990
6	3.026
7	3.054
8	3.078
10	3.113
12	3.138
100	3.269

Sector

λ.	$P\Lambda^2A^{-1}$
1/12	2.554
1/10	2.603
1/8	2.657
1/6	2.713
1/4	2.765
1/3	2.780
1/2	2.781
1	2.758

Bounded; ratio of extreme values listed: 2.000 = 2. Cf. 5.4.

$P\bar{r}^{-4}A^{-4}$

Domain	$2\pi^3 P\bar{r}^{-4}A^{-4}$
Circle	1.000
Regular hexagon	1.011
Ellipse a/b = 2	1.0125
Square	1.058
Semicircle	1.065
Equilateral triangle	1.209
Triangle 45°, 45°, 90°	1.325
Triangle 30°, 60°, 90°	1.441
Ellipse	$1 + \frac{(a-b)^4}{8ab(a^2+b^2)}$
Narrow ellipse	$\sim \frac{1}{8}\frac{a}{b} = .125\,\frac{a}{b}$
Narrow rectangle	$\sim \frac{\pi^3}{384}\frac{a}{b} = .0807\,\frac{a}{b}$
Any domain with $\pi \dot{r}\bar{r} = A$ (as triangle, rhombus, regular polygon)	$\frac{2}{\pi} P\dot{r}^{-4}$

Cf. 1.35 (d).

B

Cf. 1.20 (b)

Domain	B
Circle	$6.283 = 2\pi$
Regular hexagon	6.928
Ellipse a/b = 2	7.854
Square	8.000
Semicircle	8.792
Equilateral triangle	10.392
Triangle 45°, 45°, 90°	11.657
Triangle 30°, 60°, 90°	12.928
Ellipse	$\pi\left(\frac{a}{b} + \frac{b}{a}\right)$
Rectangle	$4\left(\frac{a}{b} + \frac{b}{a}\right)$
Circumscribed polygon	$\frac{L}{\rho} = \frac{L^2}{2A}$
Narrow sector	$\frac{4}{\gamma} + 4 + O(\gamma)$

Minimum for circle.

Regular polygon with n sides:

$$\dot{r} = \frac{\Gamma(1 - \frac{1}{n})}{2^{1-\frac{2}{n}}\Gamma(\frac{1}{2})\Gamma(\frac{1}{2} - \frac{1}{n})} L ,$$

$$\bar{r} = \frac{\Gamma(1 + \frac{1}{n})}{2^{1+\frac{2}{n}}\Gamma(\frac{1}{2})\Gamma(\frac{1}{2} + \frac{1}{n})} L .$$

Rhombus with angle $\pi\alpha$:

$$\dot{r} = \frac{\pi^{1/2}}{\Gamma(\frac{\alpha}{2})\Gamma(\frac{1-\alpha}{2})} L ,$$

$$\bar{r} = \frac{\pi^{1/2}}{8\Gamma(1 - \frac{\alpha}{2})\Gamma(\frac{1+\alpha}{2})} L$$

Triangle with angles $\pi\alpha$, $\pi\beta$, $\pi\gamma$:

$$\dot{r} = 4\pi f(\alpha)\, f(\beta)\, f(\gamma)\, \bar{R} \qquad 10)$$

where

$$f(x) = \frac{1}{\Gamma(x)} \left\{ \frac{x^x}{(1 - x)^{1-x}} \right\}^{1/2}$$

Narrow isosceles triangle with $\beta = \gamma$, $\alpha = 2\lambda \longrightarrow 0$ and side adjacent to $\pi\alpha$ of length 1:

$$\dot{r} = 4\lambda^{1+\lambda}[1 - (1 + \log 2)\lambda + O(\lambda^2)] .$$

Regular polygon, rhombus and triangle:

$$\pi\dot{r}\bar{r} = A . \qquad 11)$$

Cf. 1.23.

Annotations

1) j least positive zero of the Bessel function $J_o(x)$. Also $j = j_o$, see 5) below.

2) $E(\varepsilon) = \int_0^{\pi/2} (1 - \varepsilon^2 \sin^2 \varphi)^{1/2} \, d\varphi$.

3) Cf. G. Szegö, American Math. Monthly, vol. 57 (1950) pp. 474-478.

4) $\psi(s) = \Gamma'(s)/\Gamma(s)$. Cf. A. G. Greenhill, Messenger of Math., vol. 10 (1880) pp. 83-89.

5) j_p is the least positive zero of the Bessel function $J_p(x)$.

6) $\zeta(s) = \sum_{n=1}^{\infty} n^{-s}$.

7) G. E. Hay, Proceedings of London Math. Society, ser. 2, vol. 45 (1939) pp. 382-397.

8) G. Pólya, Quarterly of Applied Math., vol. 8 (1951) p. 386.

9) Aissen 1.

10) $\bar{R}$ denotes the radius of the circumscribed circle in the elementary sense; $\bar{R} = R$ if the triangle is acute, $\bar{R} > R$ if the triangle is obtuse. Stated Haegi 9. Cf. G. Szegö, Sitzungsberichte der Berliner Mathematischen Gesellschaft, vol. 22 (1923) pp. 38-47 and vol. 23 (1924), p. 64.

11) This follows for the regular polygon and the rhombus from the expressions for $\dot{r}$ and $\bar{r}$ (as given) which are derived from the Schwarz-Christoffel formula. For the triangle stated in Haegi 9.

BIBLIOGRAPHY

1. M. Aissen, Estimation and computation of torsional rigidity, Dissertation, 1951, Stanford University.

2. T. Carleman, Über ein Minimalproblem der mathematischen Physik, Mathematische Zeitschrift, vol. 1 (1918) pp. 208-212.

3. R. Courant, Beweis des Satzes, dass von allen homogenen Membranen gegebenen Umfanges und gegebener Spannung die kreisförmige den tiefsten Grundton gibt, Mathematische Zeitschrift, vol. 1 (1918) pp. 321-328.

4. J. B. Diaz and A. Weinstein, The torsional rigidity and variational methods, American Journal of Math., vol. 70 (1948) pp. 107-116.

5. G. Faber, Beweis, dass unter allen homogenen Membranen von gleicher Fläche und gleicher Spannung die kreisförmige den tiefsten Grundton gibt, Sitzungsberichte der Bayrischen Akademie der Wissenschaften, 1923, pp. 169-172.

6. M. Fekete, Über die Verteilung der Wurzeln bei gewissen algebraischen Gleichungen mit ganzzahligen Koeffizienten, Mathematische Zeitschrift, vol. 17 (1923) pp. 228-249.

7. J. Hadamard, Mémoire sur le problème d'analyse relatif à l'équilibre des plaques élastiques encastrées, Mémoires présentés par divers savants à l'Académie des Sciences, ser. 2, vol. 33 (1908), 128 pages.

8. H. R. Haegi, Sur le maximum du rayon intérieur, Comptes Rendus de l'Académie des Sciences, vol. 228 (1949) pp. 891-892.

9. H. R. Haegi, Extremalprobleme und Ungleichungen konformer Gebietsgrössen, Compositio Mathematica, vol. 8 (1950) pp. 81-111.

10. J. G. Herriot, Inequalities for the capacity of a lens, Duke Math. Journal, vol. 15 (1948) pp. 743-753.

11. J. G. Herriot, The polarization of a lens, Pacific Journal of Math., in print.

12. W. v. Ignatowski, Kreisscheibenkondensator, Académie des Sciences de l'URSS, Travaux de l'Institut Stekloff, II, 3 (1932) 104 pages.

13. G. Kirchhoff, Zur Theorie des Condensators, Monatsbericht der Akademie der Wissenschaften zu Berlin, 1877; Gesammelte Abhandlungen, pp. 101-117.

14. E. Krahn, Über eine von Rayleigh formulierte Minimaleigenschaft des Kreises, Mathematische Annalen, vol. 94 (1924) pp. 97-100.

15. J. C. Maxwell, A treatise on electricity and magnetism, 2nd edition, Oxford, 1881.

16. R. M. Morris, The internal problems of two-dimensional potential theory, Mathematische Annalen, vol. 116 (1939) pp. 374-400.

17. H. Poincaré, Figures d'équilibre d'une masse fluide, Paris, 1903.

18. G. Pólya, Beitrag zur Verallgemeinerung des Verzerrungssatzes auf mehrfach zusammenhängende Gebiete, 2. Mitteilung, Sitzungberichte der Preussischen Akademie der Wissenschaften, 1928, pp. 280-282.

19. G. Pólya, A minimum problem about the motion of a solid through a fluid, National Academy of Sciences, vol. 33 (1947) pp. 218-221.

20. G. Pólya, Sur la fréquence fondamentale des membranes vibrantes et la résistance élastique des tiges à la torsion, Comptes Rendus de l'Académie des Sciences, vol. 225 (1947) pp. 346-348.

21. G. Pólya, Torsional rigidity, principal frequency, electrostatic capacity and symmetrization, Quarterly of Applied Math., vol. 6 (1948) pp. 267-277.

22. G. Pólya, Sur la symétrisation circulaire, Comptes Rendus de l'Académie des Sciences, vol. 230 (1950) pp. 25-27.

23. G. Pólya and G. Szegö, Über den transfiniten Durchmesser (Kapazitätskonstante) von ebenen und räumlichen Punktmengen, Journal für die reine und angewandte Math., vol. 165 (1931) pp. 4-49.

24. G. Pólya and G. Szegö, Inequalities for the capacity of a condenser, American Journal of Math., vol. 67 (1945) pp. 1-32.

25. G. Pólya and A. Weinstein, On the torsional rigidity of multiply connected cross-sections, Annals of Math., vol. 52 (1950) pp. 154-163.

26. Lord Rayleigh, The theory of sound, 2nd edition, London, 1894/96.

27. B. de Saint-Venant, Mémoire sur la torsion des prismes, Mémoires présentés par divers savants à l'Académie des Sciences, vol. 14 (1856) pp. 233-560.

28. M. Schiffer and G. Szegö, Virtual mass and polarization, Transactions of the American Math. Society, vol. 67 (1949) pp. 130-205.

29. G. Szegö, Verallgemeinerung des ersten Bieberbachschen Flächensatzes auf mehrfach zusammenhängende Gebiete, Sitzungsberichte der Preussischen Akademie der Wissenschaften, 1928, pp. 477-481.

30. G. Szegö, Über einige Extremalaufgaben der Potentialtheorie, Mathematische Zeitschrift, vol. 31 (1930) pp. 583-593.

31. G. Szegö, Über einige neue Extremaleigenschaften der Kugel, Mathematische Zeitschrift, vol. 33 (1931) pp. 419-425.

32. G. Szegö, On the capacity of a condenser, Bulletin of the American Math. Society, vol. 51 (1945) pp. 325-350.

33. G. Szegö, The virtual mass of nearly spherical solids, Duke Math. Journal, vol. 16 (1949) pp. 209-223.

34. G. Szegö, On membranes and plates, National Academy of Sciences, vol. 36 (1950) pp. 210-216.

35. G. Szegö, Über eine Verallgemeinerung des Dirichletschen Integrals, Mathematische Zeitschrift, vol. 52 (1950) pp. 676-685.

36. G. Szegö, Conformal mapping related to torsional rigidity, principal frequency and electrostatic capacity, The construction and applications of conformal maps: Proceedings of a Symposium held at the Institute for Numerical Analysis, Los Angeles, June 1949, in print.

37. G. I. Taylor, The energy of a body moving in an infinite fluid, with an application to airships, Proceedings of the Royal Society, London, ser. A, vol. 120 (1928) pp. 13-21.

38. A. Weinstein, Étude des spectres des équations aux dérivées partielles de la théorie des plaques élastiques, Mémorial des Sciences Mathématiques, vol. 88 (1937), 62 pages.

LIST OF SYMBOLS

On account of the unusually large number of symbols used in this book, it seemed inevitable that occasionally the same symbol was employed for two different purposes. We tried to eliminate the danger of confusion by avoiding such occurrences in the same context.

Limitations of a technical nature compelled us to introduce certain simplifications in the notation. For instance, P is used for the torsional rigidity instead of the Greek capital Rho; underlined capitals such as $\underline{A}$, $\underline{B}$, . . . are used in lieu of script.

$\underline{A}$	closed surface	1
$\underline{B}$	solid	1
$\underline{C}$	closed curve in a plane . .	1
$\underline{D}$	domain in a plane	1
V	volume	1
S	surface area	1
C	electrostatic capacity; later conductance, p. 49 . .	1
M	integral of mean curvature .	1
W_m	average virtual mass	1, 240
P_m	average polarization	1, 240
A	area	1
L	length	2
I_a	polar moment of inertia (with respect to a)	2
I	polar moment of inertia (with respect to centroid) .	2
r_a	inner radius (with respect to a)	2
$\dot{r}$	maximum inner radius	2
$\bar{r}$	outer radius	2
P	torsional rigidity	2
$\Lambda = \Lambda_1$	principal frequency of membrane	2
$j=2.4048$	first positive root of the Bessel function $J_0(x)$	3
$h=3.1962$	first positive root of $J_0(x)I_0'(x)-J_0'(x)I_0(x)$	31
$k=3.8317$	first positive root of $J_0'(x)=-J_1(x)$	31
$\bar{V}$	volume-radius	3
$\bar{S}$	surface-radius	3
$\bar{M}$	mean radius	3
$\bar{A}$	area-radius	4
$\bar{L}$	perimeter-radius	4
$\bar{I}$	polar moment-radius . . .	4
$\bar{C}$	capacity radius of plate	4
$\bar{P}$	torsional rigidity-radius	4
$\bar{\Lambda} = \bar{\Lambda}_1$	principal frequency-radius	4
R_1, R_2	principal radii of curvature	14
$B_a = \int h^{-1}ds$	h distance of tangent from a	14
$B = \min B_a$	. . . .	14

Lightning Source UK Ltd.
Milton Keynes UK
UKHW030659290123
416095UK00001B/1

9 780691 079882